New Media in Europe
Satellites, Cable, VCRs and Videotex

New Media in Europe

Satellites, Cable, VCRs and Videotex

John Tydeman

Ellen Jakes Kelm

McGRAW-HILL Book Company (UK) Limited

London · New York · St Louis · San Francisco · Auckland
Bogotá · Guatemala · Hamburg · Johannesburg · Lisbon · Madrid
Mexico · Montreal · New Delhi · Panama · Paris · San Juan
São Paulo · Singapore · Sydney · Tokyo · Toronto

Published by
McGRAW-HILL Book Company (UK) Limited

MAIDENHEAD · BERKSHIRE · ENGLAND

British Library Cataloguing in Publication Data
Tydeman, John
  New media in Europe: Satellites, Cable, VCRs and Videotex.
  1. Mass media—Europe
  I. Title   II. Kelm, Ellen Jakes
  302.2′34   P92.E9

ISBN 0-07-084799-1

Library of Congress Cataloging in Publication Data
Tydeman, John.
  New media in Europe.
  Includes bibliographies and index.
  1. Television broadcasting—Europe. 2. Television
advertising—Europe. 3. Moving-picture industry—Europe. 4. Video tape recording
industry—Europe. 5. Telecommunication—Europe. I. Kelm, Ellen Jakes.
II. Title.
HE8700.9.E8T93   1986   384.55′094   85–19839
ISBN 0-07-084799-1

12345 B&T 89876

Printed and bound in Great Britain by Butler & Tanner Limited, Frome, Somerset

Contents

Preface

The successful launch of Eutelsat-F1 in June 1983 and the commencement of pan-European cable services marked a new era in mass market media in Europe. What had become possible via Westar in 1975 in the United States—national cable networks—was now a reality for all European cable households in 1983. Cable could become more than a broadcast relay service.

The European coverage offered by Eutelsat raised hopes that pan-European advertising may provide outlets for advertisers not readily available on the national television networks.

To add to the excitement, national governments had made commitments to direct broadcast satellite systems which were to appear by 1985. However, there was neither the infrastructure, the support industry, nor national government policies in place to capitalize on this opportunity.

The promise still exists, as the market opportunity has not changed significantly, but implementation is slower and more difficult than was first anticipated.

This book has been written for those who are interested or involved in the European new media market. Our intention has been to represent the state of new media in Europe by late 1985, to assess the present developments qualitatively and quantitatively, and, finally, to offer a basis for evaluating these opportunities. As such, the book is written for those inside and outside Europe who want to be informed of developments in the field of consumer electronic media in Europe.

In Chapter 1, we define Europe as a market and make brief comments about pertinent dimensions of that market for the subsequent discussion. It is recognized that Europe is not a single market—historically, politically, culturally, or economically—despite the European Economic Commission (EEC). Still, it is an 'accepted' bloc of countries. Throughout the book, we focus on the differences as much as the similarities.

Mass market electronic media in the eighties is evolving from the television set. Cable, satellite, and video recorders are ways of bringing entertainment and information to the household. They can exist only because there is a television set in 95% of households.

To describe the present position in Europe, we have reviewed the structure and operation of television broadcasting (Chapter 2), outlined the situation for advertising, particularly television advertising, and the potential for advertising growth through cable and satellite (Chapter 3), and described the cinema industry—an industry which is already being impacted by video cassette recorders (Chapter 4).

We assess the satellite position in Europe for cable relay and direct broadcast and estimate the likely market potential for DBS (Chapter 5), review cable and master antenna systems, their market growth and political obstacles (Chapter 6), and describe the video cassette and video disc markets (Chapter 7). These three chapters focus on what could be termed the delivery options for the household, rather than the specific contents of new media.

In Chapters 8 and 9, we turn to content—entertainment services in operation or proposed for cable and DBS (Chapter 8) and information services being provided via the television receiver, teletext, and videotex (Chapter 9).

Finally, we have attempted to identify the political, technical, and regulatory uncertainties against which forecasts of new media growth are usually made, and have organized these to help readers monitor and track the developments in Europe (Chapter 10). Our aim throughout has been to provide a framework in which to monitor change in new media.

The opinions in this book are solely those of the authors and we alone are responsible for all errors and omissions. We wish to thank Rowena Lord and Michael Flint for their helpful comments on the regulatory issues in Europe and to Candace Johnson for her critique of the sections on cable and satellite developments in Central Europe.

Finally, we are indebted to Hiroshi Iwai of Mediahouse, Inc., who persisted for so long that we finally agreed to write the book.

1

Europe in perspective

1.1 Introduction

Until the late seventies, the consumer market in Europe for entertainment and information was relatively ordered. The broadcast industries were heavily regulated and dominated by the state, national press was controlled by a small number of media giants, and cinemas aligned themselves into major chains. Only regional publications were small and relatively unregulated.

The eighties have seen a change in this relationship. The so-called 'new media' or 'electronic media' are infiltrating the European households from all directions. The television set which has achieved mass market penetration in every country is at the core of this consumer media revolution.

Black and white consoles have already given way to modular-designed colour television with remote-control keypads and stereophonic sound (or simulcast). Large screens, flat screens, and very large-scale integrated chip sets to improve performance are coming and, in the not-too-distant future, digital transmission and high-definition television will be available. Television sets come with teletext and videotex capabilities. They are used with personal computers to provide home information centres. Video cassette recorders provide a home cinema and time-shifting of broadcast programmes using the television set put entertainment further under the control of the user). Cable and direct broadcast satellites will increase the choice of television entertainment as well as provide the potential for at-home information, education, and business services.

This media development is not just a European phenomenon; rather, it is occurring world-wide and is changing the balance of power in the broadcast and print industries as well as providing opportunities for existing and new corporations to invest and diversify. The economic and social importance continues to increase. A variety of market forecasts for the various industries predicts that there are potentially billions of dollars at stake. In the United States, for example, studies show that nearly 50% of the workforce is already engaged in entertainment and information-related

1

activities and that an even greater percentage share of the Gross National Product (GNP) is derived from these industries.

1.2 Overview

We have structured our analysis of new media in Europe into four sections. The first section is a brief review of traditional electronic media—television, radio, and cinema—highlighting national differences and similarities in:

—Organizational structure
—Regulations
—Revenue, funding, and costs
—Penetration and usage

Special emphasis is given to the importance of advertising in the evolution of broadcasting. It is the potential for additional advertising which is critical in the assessment of the viability of most proposed cable or satellite-delivered channels.

The second and third sections review new electronic media—its delivery to the home, its potential market, and its content. There are three new delivery options for linking the consumer household to the source of entertainment and information. First, there is a national and even international link via satellite from the source of programming directly to the household—direct broadcast satellite—or to a cable system. This, more than any other delivery mechanism, promises to change the existing national broadcast infrastructure.

Second, there is a physical hardware link—coaxial or fibre optic cable. Cable is usually a local system which is fed by microwave, satellite, and terrestrial broadcast transmissions. Cable can allow for two-way interactivity as well as providing information and telephone services.

Third, there is the home-based analogue of the cinema in the eighties: the video cassette recorder. Consumers buy, rent, or 'copy' (tape) material to be viewed under their control at home. This is a very different, but equally important, delivery option of entertainment and information to the household and is not necessarily a substitute for cable or satellite.

Delivery options, while important, are only one aspect of the new electronic media. The content (or, in McLuhan terms, the message) is more important than the transmission medium. In the third section, the analysis of new services is directed primarily at the newly emerging satellite and terrestrial entertainment channels for Europe. We also review the developments of electronically delivered information services to the home, primarily videotex and teletext, to complement the discussion on entertainment services and to indicate the extensiveness of the evolving new media.

In the fourth section, we provide a context for new media market development by considering the existing and emerging regulatory environment for new pan-European television channels. The key regulatory variables are identified in the overall process.

Throughout the book, we have expressed most expenditures and revenues in US dollars—a common, relatively accepted currency measure. The surge of the dollar against European currencies in late 1984 and its fall in mid-1985 means that, while comparisons retain a relative significance, the absolute dollar amounts need careful interpretation.

1.3   A formal definition of Europe

Our focus is the consumer markets of Europe, particularly Western Europe, an economic bloc of around 350 million people from 16 major nations (Table 1.1) speaking at least six languages—Italian, English, German, French, Spanish, and Swedish—as well as the other national languages such as Flemish, Finnish, Greek, Portuguese, and numerous dialects. The European Economic Community is evidence of the perceived benefits to be gained for Europeans to operate as a single entity. Yet national differences far outweigh national similarities. Even the notion of geographical proximity is misleading. The tip of Norway to Gibraltar in the Mediterranean is about the same distance as from Singapore to Tokyo or Mexico City to Newfoundland.

Formally, we define Europe to be the nations of Scandinavia and Western and Southern Europe. This includes all members of the European Economic Community (EEC) and major members of the Council of Europe (CE):

Country	Membership
Denmark	EEC, CE
Finland	Observer CE
Norway	CE
Sweden	CE
Austria	CE
West Germany	EEC, CE
Switzerland	CE
Belgium and Luxembourg	EEC, CE
Netherlands	EEC, CE
France	EEC, CE
Italy	EEC, CE
United Kingdom	EEC, CE
Ireland	EEC, CE
Spain	EEC, CE
Portugal	EEC, CE
Greece	EEC, CE

Table 1.1 Profile of Europe

Country	Population (million)	Households (million)*	Household size*	GDP ($ billion)†	GDP per head ($)‡	Imports as a percentage of GDP 1983	Household expenditure for recreation, entertainment, education, and cultural services as a percentage of household consumption	Television penetration of all households (%)¶	Colour television penetration of all households (%)	Amount of television viewed per adult per day**
Austria	7.5	2.7	2.8	67.1	8 947	29.11	5.9	93	64	2 h 18 min
Belgium	10.0	3.3	3.0	81.9	8 190	66.85	5.9	96	71	2 h 55 min
Denmark	5.2	2.2	2.4	56.5	10 858	29.90	9.1	90	67	1 h 54 min
Finland	4.9	1.9	2.6	49.5	10 096	27.63	8.1	97	56	2 h
France	54.7	19.3	2.8	516.4	9 441	21.40	6.5	93	56	2 h 9 min
Greece	10.1	2.8	3.6	34.5	3 416	26.59	4.2	93	14	3 h
Ireland	3.6	0.9	4.0	18.1	5 025	55.34	9.2	92	60	2 h 19 min
Italy	58.5	18.4	3.2	352.8	6 031	24.92	7.5	96	43	2 h 40 min
Netherlands	14.6	4.8	3.0	132.2	9 055	45.61	9.8	97	82	1 h 27 min
Norway	4.2	1.4	3.0	55.0	13 095	27.55	9.8	93	79	1 h 18 min
Portugal	10.5	3.0	3.5	20.7	1 969	40.43	N/A	90	14	3 h
Spain	38.9	10.5	3.7	158.7	4 080	17.65	7.0	94	42	3 h
Sweden	8.4	3.3	2.5	91.7	10 916	28.22	10.4	97	60	2 h
Switzerland	6.4	2.3	2.8	97.1	15 170	29.82	9.6	85	69	1 h 34 min
United Kingdom	56.2	20.8	2.7	456.5	8 123	21.15	8.1	98	77	3 h 10 min
West Germany	61.9	25.1	2.5	653.9	10 563	23.36	10.4	98	80	2 h 13 min
Total	355.6	122.7	2.9	2842.6	7 993					

* From Euromonitor Publications Ltd.[1]
† From Bureau of Statistics of the International Monetary Fund.[2]
‡ From OECD.[3]
§ From European Economic Community.[4]
¶ From Campaign and Young and Rubicam.[5]
** From J. Walter Thompson Europe.[6]

4

Comparable data are *not* always available for all countries; nor are all new media developments taking place in every country. A number of smaller nations have been omitted from discussion in this book. They include Andorra, Cyprus, Iceland, Liechtenstein, Malta, Monaco, San Marino, Turkey, and the Vatican.

Europe in its entirety can provide a market that is larger than either the United States or Japan. At the same time, individual countries differ economically, and entertain extremely diverse policies, regulations, attitudes, and cultural values to existing media and new media.

1.4 Demographics

Europe is the largest market in the Western world. It has over 350 million inhabitants, well above the population of the United States which has 230 million or Japan which has 118 million. Europe also covers a greater geographic area, but is only half as densely populated as Japan.

Language and culture divide Europe into distinct segments, with over 13 languages and many other dialects. The major language groupings by national population are: German (20.7%), French (16.8%), English (16.9%), Italian (16.7%), Spanish and Portuguese (13.9%), the rest (Greek, Dutch and Flemish) (8.7%). Scandinavia, a bloc of 22.7 million people, is divided

Table 1.2 European highlights

Household size			GDP per head		Household expenditure for recreation, entertainment, education, and cultural services as a percentage of the total final household consumption	Percentage of colour television penetration
Low	1	United Kingdom	High	Switzerland	Sweden	Netherlands
	2	France		Norway	United Kingdom	West Germany
	3	West Germany		Sweden	Switzerland	Norway
	4	Sweden		Denmark	Ireland	United Kingdom
	5	Denmark		Finland	Denmark	Belgium
High	12	Italy	Low	Italy	Spain	France
	13	Portugal		Ireland	France	Finland
	14	Greece		Spain	Belgium	Norway
	15	Spain		Greece	Austria	Portugal
	16	Ireland		Portugal	Greece	Greece

among several similar, but distinct, languages, accounting for 6.3%. However, more people speak English than any other language throughout Europe.[7]

Population in Europe is not expected to rise dramatically over the next decade or so, and negative growth has been evidenced in some of the Scandinavian countries, West Germany, and the United Kingdom.

Europe is experiencing a similar trend to the United States of a reduction in household size as more couples and individuals move into their own dwelling units. Consequently, the number of households is increasing and the average household size of 2.9 people (see Table 1.1) is falling. Greece, Ireland, Portugal, and Spain are all significantly above the average, while West Germany, Sweden, and Denmark are below (see Tables 1.1 and 1.2).

As media consumption is primarily measured by households, this trend will affect the size of the potential market over the next decade. At present, there are 123 million households (see Table 1.1), compared to 82 million in the United States.[8]

1.5 Economics
Economically, Europe presents a weaker front than either the United States or Japan. With a strong global dollar and yen, Europe has suffered economic setbacks.

In 1983 and 1984, real growth in the gross national product (GNP) in the United States and Japan was considerably higher than OECD Europe. In the United States, the growth was 3.4 and 6.0% in 1983 and 1984 and in Japan, the corresponding figures were 3.0 and 4.75%. In Europe, the growth was 1.3 and 2.25%. The growth rate is forecast to slow down in both the United States and Japan, but not to increase appreciably in Europe. Total domestic demand in Europe is projected to grow at 2%, increasing from the low growth of 1.0 and 1.75% in 1983 and 1984. Again, growth in the United States and Japan has been higher at 7.25 and 3.75% in 1984 respectively and is forecast to grow at a greater rate than in Europe at 3% in the United States and 3.25% in Japan.[9]

The differences in relative individual wealth within European countries, unadjusted for taxation, are enormous. At one end, Denmark, Finland, Norway, Sweden, Switzerland, and West Germany all have a per capita gross domestic product (GDP) in excess of $10000, while the per capita GDP in Greece, Portugal, and Spain is less than $5000 (Table 1.1).

While Japan has experienced low levels of inflation over the past few years, both the United States and Europe have experienced escalating prices. The United States has reduced the growth rate of consumer prices to 3 to 4%, but Europe, as a whole, is still increasing at a 6 to 7% rate. These rates vary considerably among European countries.

6

The level of unemployment in Europe has been higher than either the United States or Japan, and this trend is likely to continue. From 12 to 15 million people are expected to be unemployed, accounting for over 10% of those employed.[9] Unemployment has decreased in the United States and remained low in Japan, but this remains one of the most serious problems to be faced in Europe, and could inhibit economic growth.

Business confidence throughout Europe is expected to increase, and a halt in the decline of corporate profits has been forecast. The overall economic situation in Europe is not nearly as positive as in the United States or Japan, but it appears to be improving, and should be more stable in the late eighties.

The Europeans on the whole do not have the affluent life style of the Americans, although there are several countries where social and personal conditions are higher by standard measures. For example, consider the share of final household consumption allocated to recreation, entertainment, education, and cultural services. While no countries reach the *percentage allocations* comparable to the United States (13%), those countries with a higher relative GDP per capita, such as Switzerland, Sweden, and Denmark, also devote a larger share of their incomes to leisure pursuits (Tables 1.1 and 1.2). It is apparent that there is excess income and time to enjoy leisure pursuits in nearly every European country.

1.6 Trade and industry

Within the EEC, trade among countries is very important. There are two major agreements in place which have particular impact upon the electronics industry: Articles 59 and 60 in the *Treaty of Rome*.[10] They state that: 'Restrictions on freedom to provide services within the Community will be progressively abolished' where services are defined to include:

1. Activities of an industrial nature
2. Activities of a commercial character
3. Activities of craftsmen
4. Activities of the professions

Under this agreement, the members of the EEC countries are able to trade freely. Media and information services are included under the definition of services, although they were not specifically mentioned in the *Treaty*.

The total imports and exports of Europe exceed those of the United States and Japan combined. This is due in part to the small size of the individual countries, and their inability to supply their own domestic needs. Consequently, they trade extensively among one another. Imports, for example, exceed 40% of GDP in Belgium, Ireland, Netherlands, and Portugal,

7

are between 25 and 40% in Austria, Denmark, Finland, Greece, Norway, Sweden, and Switzerland, and are between 15 and 25% for France, Italy, Spain, the United Kingdom and West Germany (see Table 1.1).[11]

The market for foreign-produced media hardware in Europe is large because Europe, as a whole, does not produce much of its consumer electronics products domestically. Europe is very reliant on exports from Japan and the United States, and this has increased as media consumer durables have become more technological in nature. The EEC subscribes to the GATT agreements and the Tokyo round of discussions has been particularly relevant because it has imposed quotas on the volume of Japanese hardware which enters Europe. This agreement was enacted to protect domestic European brands, particularly the V2000 VCR developed by Philips-Grundig. Imports into France were further slowed by limiting entry for electronic products to just one port.

At present, public research and development expenditure in Europe is lower than in the United States or Japan, at about 1% of total GDP or $26 billion. Further, Europeans spend, on average, 25% more than American or Japanese companies in the transformation of research and development to a finished product. Consequently, similar products are priced higher, and Europe is not as competitive on products that are closely tied to research and development: high technology and new media.[12]

In an effort to improve the situation, a project called ESPRIT has been developed in the EEC to stimulate the high technology industries in Europe. In this project 750 million European community units (ECUs) have been allocated for a cooperative research and development effort, encouraging Europeans to develop their own industry. Many analysts believe that a good portion of the capital spent on research efforts will be devoted to catching up, rather than advancement, and the United States and Japan will move even further ahead. Those companies involved in ESPRIT are:[12]

Philips	Netherlands
Siemens	Germany
GEC	United Kingdom
ICL	United Kingdom
Plessey	United Kingdom
Thomson	France
CIT-Alcatel	France
CU Honeywell/Bull	France
Olivetti	Italy

A follow-on initiative is the Eureka project in which four of the companies, GEC, Thomson, Philips, and Siemens, have agreed to spend around

$5 billion over 10 years to develop advanced microprocessors, high-density memories, flat screens, and integrated circuits.[13]

At the same time, there is an antitechnology sentiment by many workers in Europe who feel new technology is threatening jobs. As a result, the introduction of new technologies, such as computers for use in newspaper publishing, has been met with protest.

Because Europe is not developing its new technology industries as quickly as the United States and Japan, it is becoming more dependent on outside sources of technology. The lack of internal exposure within the countries is also leading to a slower acceptance rate.

1.7 Telecommunications and media

In contrast to other areas, Europe has managed to maintain control over its telecommunications industries, meeting two-thirds of domestic demand internally. This is because the state-controlled PTTs (postal, telegraph, and telecommunications departments) are the largest single purchasers and they protect their own telecommunications markets through direct support.

Despite this, barriers in this area are eroding for several reasons:

1. Changes in the industrial environment over the last few years
2. Satellite services are growing and are harder to control than land-based transmissions
3. Costs of new switching systems are straining even the large resources of the PTTs

As a result, private industry is more likely to break into this near monopoly in Europe, and the internal hold on telecommunications could begin to diminish. In the United Kingdom, Mercury Communications Ltd has been set up by the Government with a mandate to compete with British Telecom, which was itself 'privatized' in 1984.

It is apparent that, in general, the EEC companies must look at the European market as a whole to prevent further fractionalization of industry. Electronics companies have developed numerous factories in each individual country which encompass all aspects of manufacture of a specific product. The individual factories are not always large enough to support the merit of such a base. The lack of specialization, which is more efficient, has made it difficult for Europe to compete with Japan or the United States, who approach Europe as an entire market and develop their specialized plants accordingly. Ultimately, commercial self-interest and economic pressures must lead to industrial rapprochement in each country.

Telecommunications is only one story. Cooperation is also necessary for broadcasting systems to function, and Europe has managed to devise a working relationship among the countries. Broadcasting is primarily under governmental control. There are relatively few radio and television stations, with the exception of Italy, and most systems operate nationally, although some regional stations do exist.

Commercialism on broadcast media is much less than in the United States, and many European stations are supported by licence fees levied on the television receiver. In addition, rentals, rather than outright purchase of television and more recently, VCRs, is not uncommon.

Advertising patterns vary considerably throughout Europe, but they are particularly different from Japan or the United States. The broadcast media are more restricted than they are in either the United States or Japan. As a result, the media mix for advertisers is radically altered, and there is greater expenditure for the print media than television. In addition, cinema advertising is widespread, an outlet that rarely exists in the United States. With programming not directed at mass audiences, there is a large demand for advertising on television that is unfulfilled.

Europe is also divided on some strategic technological standards, a problem that does not arise in a homogeneous country like the United States or Japan. Media transference in both hardware and software has been particularly beset by divisions, with varying transmission standards for television, VCR formats, videotex, and satellite systems.

While some cooperative organizations have been formed, such as the European Broadcasting Union (EBU) and European Satellite Association (ESA), the conflict among political parties and their respective policies towards new media has prevented the equilibrium that exists in the United States or Japan. However, it appears that Europe has realized that it must confront the technology advance, and policy making in this area has increased, with signs of further cooperation.

1.8 Television viewing

Three activities dominate European leisure time: television viewing and listening to the radio, social contacts with friends and family, and reading. Television generally takes up 25 to 33% of total leisure time and nearly 20% of Europeans rate it as their chief leisure pursuit. Television viewing ranks as the major leisure pursuit for all except two countries: Germany and Austria.[14]

Both television and colour television penetration is high. Only Switzerland has less than 90% of household penetration of television and only Greece, Italy, Portugal, and Spain have less than 50% of households with colour television (Table 1.1).

The market for home electronics is expanding and is growing at three times the rate of the leisure market. There has been a significant growth in the acquisition of colour television sets and video equipment. Per capita spending for the electronics market is highest in Denmark and lowest in Finland.[15]

Despite the fact that the television market is thriving, there is general dissatisfaction with the services and programmes offered by most European broadcasting systems. Poor quality of programming and dissatisfaction with programme type are the most common reasons given for negative assessment by viewers. Viewers feel that there are too many programmes with discussion, public service, and political content, while there are too few entertainment programmes, such as films, plays, and sports. Close governmental control is also responsible for considerable viewer dissatisfaction, particularly in France, Greece, Spain, and West Germany.[6]

Viewers in Austria, Finland, Ireland, and the United Kingdom are relatively satisfied with their choice of services and Belgium and the Netherlands are neither negative nor positive. However, it is apparent from the average hours viewed each evening by adults that programming does not have the same entertainment value as it does in the United States, where viewing stands at 4 hours 12 minutes per day.

United Kingdom adults watch the most television per day at 3 hours 10 minutes, followed by Spain, Greece, and Portugal at 3 hours. Interestingly enough, these countries are among the most commercial in Europe; the United Kingdom has the most advertising and the other three countries have their television broadcasting funded almost entirely through advertising. At the other end of the spectrum, the highly educational channels of Norway and the Netherlands are the least watched (Table 1.1).

1.9 Conclusion

Europe is the largest market in the Western world, both in size and population. It has more inhabitants per household, which gives each home a larger potential of viewers and users of media services. However, the average GDP per capita distributed throughout Europe ($7993) is lower than the United States ($13 857—based upon a population of 235 million) and Japan ($9657—based upon a population of 120 million), and Europe has experienced less economic growth than these countries, as well as suffering a higher rate of unemployment.

The GDP of most European nations contains a high percentage of both imports and exports, as considerable trading is done among the various nations for all products. Europe is particularly open to exports of media hardware from outside, and it has been forced to impose quotas to allow its own industries to survive. The establishment of the ESPRIT programme by the EEC is an attempt to encourage domestic development in electronics, and many of the major European companies in this industry are involved.

Dissension between governments in Europe occurs over many issues, with the result that standards and policies are often incompatible. This is particularly true for media, transmission standards, and advertising policies, a situation which is beginning to change as Europeans attempt to fend off Japanese and American products and services. However, the diversity of opinion means that each new policy will probably take longer to implement than it would in the United States or Japan because each national faction will want to influence the outcome.

The broadcast media are primarily under the control of the national governments, and their advertising is restricted. Satisfaction with television is low and the consumer market is ready for new and varied sources of entertainment programming.

Our aim has been to 'set the scene' for an evaluation of new consumer media throughout Europe. Grouping countries according to potential for new media is not without risk. Ignoring national media regulations at this stage (discussed in Chapters 2 and 10), it is possible to hypothesize that receptivity of new media is a function of economic wealth (GDP per head, household size), the propensity to spend on recreation and entertainment services (expenditure as a share of total household consumption), and the ability to receive new media (colour television penetration) (Table 1.2). By simply ranking countries according to the above criteria, we obtain fairly distinct groupings. Those which have the greater potential for new media include Sweden, the United Kingdom, and Denmark, while those with the least economic potential would appear to be Greece, Portugal, and Spain.

The diversity of media in Europe ensures that the situation for new media will be different to that of Japan and the United States. It remains to be seen whether new media paths within individual countries will be similar, or if European nations will embark on entirely separate paths during the remainder of the eighties.

References

1. Euromonitor Publications Limited, *European Marketing Data and Statistics 1982*, Euromonitor Publications Limited, London, 1982.
2. Bureau of Statistics of the International Monetary Fund, *International Financial Statistics*, Volume XXXVIII, International Monetary Fund, Washington, DC, 1985.
3. The Organization for Economic Co-operation and Development, *OECD Economic Outlook*, No. 35, OECD, Paris, 1984.
4. European Economic Community, *Eurostatistics*, European Economic Community, Brussels, 1983.
5. Campaign and Young and Rubicam, *The Advertisers Guide to European Media*, Marketing Publications Ltd, London, 1983.
6. J. Walter Thompson Europe, *Television Today and Television Tomorrow*, 7. Walter Thompson Co. Ltd, London, 1983.

7. Advertising Association, *Marketing Pocket Book 1984*, The Advertising Association, London, 1984, pp. 110–111.
8. A. C. Nielsen Company, *'84 Nielsen Report on Television*, A. C. Nielsen Company, Northbrook, Illinois, 1984.
9. The Organization for Economic Co-operation and Development, *OECD Economic Outlook*, No. 35; quoted in 'At best the outlook for jobs is bad, at worst it's awful', *The Economist*, 23 June 1984, pp. 63–64.
10. European Communities, *Treaties Establishing the European Communities*, Articles 59 and 60, Office for Official Publications of the European Communities, the Netherlands, 1978, pp. 268–269.
11. The Organization for Economic Co-operation and Development, *The United States*, Organization for Economic Co-operation and Development, Paris, 1983–84.
12. Guy de Jonquieres, 'Collaboration: now Europe tries again', *The Financial Times*, 21 March 1983, p. 11.
13. 'Four electronics firms set to cooperate on Eureka', *Wall Street Journal*, 20 June 1985, p. 3.
14. William H. Martin and Sandra Mason, *Leisure Markets in Europe*, Volumes I, II, and III, Financial Times Business Publishing Division, London, 1978.
15. Mackintosh Consultants Ltd, *Mackintosh Yearbook of West European Electronics Data 1984*, Mackintosh Consultants Ltd, London, 1984.

2

Television broadcasting

2.1 Overview

Throughout most of Europe, a broadcast authority has been granted a licence by the government to offer the national television services for a monopoly on the airwaves. The subsequent degree of government control or influence over content varies from virtually nil, such as in the United States, to extensive control, such as in Norway. Channels are supported through a combination of government funds which are generally obtained through licence fees levied on the receiver sets and some advertising revenues. Originally, there was no commercial broadcasting. Increasing costs, changing attitudes, and viewer demands opened the way for the introduction of limited commercial airtime in some countries.

The British Broadcasting Corporation (BBC) was the first European broadcasting organization to be established in the twenties. Other European countries modelled their own broadcasting infrastructures on the BBC, to 'educate, entertain and inform'.[1] A large percentage of programming is devoted to news, consumer affairs, and educational programmes, much more than the more entertainment-oriented fare provided on US television. Local programming content is relatively expensive as it is provided for by a highly unionized industry and a broadcaster with guaranteed income through licence fees.

The content dilemma is clear: programming to fulfil a broadcast authority's obligation to transmit and support cultural, informative, and educational programming versus programming to satisfy the popular demands of the consumers, i.e., the licence-fee-paying public. Ratings analyses leave no doubt as to what is demanded in the market—popular entertainment-oriented programming.

Nations have responded differently to the dilemma. In Norway, for example, the entertainment dimension is completely overlooked, so much so that the top-ranked US soap opera, *Dallas*, was banned from the national television and became one of the most widely sold video tapes. On the other hand, the BBC has been 'accused' of competing with the commercial

14

broadcasters in the United Kingdom for high ratings. Its 1985 request for a substantial increase in licence fees brought into question the role that the BBC should play in UK broadcasting and how it should be funded.

Before examining the new and emerging cable and satellite channels (Chapter 8), we review the current state of television broadcasting: its organization, regulations, sources of revenue, programme content, and future developments. No industry concedes ground to competitors without a fight, and the broadcasters of Europe are no exception.

2.2 European Broadcasting Union

The European community encourages programme cooperation and exchange through the European Broadcasting Union (EBU). The EBU was formed at the instigation of the BBC in 1950, following the break-up of the International Broadcasting Union after the Second World War. The Soviet bloc formed its own organization, the International Broadcasting Organization (now the OIRT). The remaining European countries formed the EBU, which has its headquarters in Geneva, Switzerland, and its technical centre in Brussels, Belgium. Over 32 nations and 40 broadcast authorities are members. Associate worldwide members add a further 69 broadcasters.[2] The member countries and organizations are listed in Table 2.1.

A broadcasting organization must be a member of the International Telecommunications Union (ITU) in order to become an EBU member. In addition, it must produce and have authority over programmes, operate at least one permanent transmitter, and provide a national service.

A 15-member administrative council acts on behalf of the total organization. Switzerland is always accorded a seat. The other 14 seats are essentially open to nominations, but the five countries which contribute the heaviest subscription fees are also always represented: the United Kingdom, West Germany, Italy, France, and Spain. The other members rotate membership for the remaining nine seats. Four major committees oversee the EBU activities: technical, legal, radio programmes, and television. Within these committees, the EBU unites television programmes for programme exchange, setting of tariffs, establishing and monitoring legal contracts and technical standards, as well as transborder control of programmes to prevent violation of national programme rules.[4]

The EBU also collaborates on programming, particularly news and special events, and often shares production facilities and rights to major events, such as the Olympics, which reduces costs to the individual country and increases the international leverage of Europe as a whole for negotiation of television rights to international events. For example, by collectively bargaining for the Olympic rights with the International Olympics Committee for the 1984 Olympics, the EBU paid only $19.8 million to reach 120

Table 2.1 EBU member countries and organizations[3]

Country	Organization
Algeria	Radiodiffusion-Télévision-Algérienne
Austria	Österreichischer Rundfunk
Belgium	Belgische Radio en Televisie, Nederlandse uitzendingen
	Radio-Télévision Belge de la Communauté Française
Cyprus	Cyprus Broadcasting Corporation
Denmark	Radio Danmarks
Finland	Oy Yleisradio Ab
France	Organismes Française de Radio-télévision Europe No. 1
	Télécompagnie
Greece	Elliniki Radiophonia Tileorassis
Ireland	Radio Telefis Eireanne
Israel	Israel Broadcasting Authority
Italy	RAI—Radiotelevisione Italiana
Jordan	Jordan Television
Lebanon	Radiodiffusion Libanaise/Télé-Liban
Libya	Libyan Jamaheuya Broadcasting
Luxembourg	Radio Télé-Luxembourg
Malta	Broadcasting Authority Malta/Xandi Malta
Monaco	Radio Monte Carlo
Morocco	Radiodiffusion-Télévision Marocaine
Netherlands	Netherlands Omroep Stichtine
Norway	Norsk Rikskringkastring
Portugal	Radiodifusao Portuguesa EP
	Radiotelevisao Portuguesa EP
Spain	Radiotelevision Española
	Sociedad Española de Radio difusion
Sweden	Sveriges Radio AB
Switzerland	Société Suisse de Radiodiffusion et Télévision
Tunisia	Radiodiffusion-Télévision Tunisienne
Turkey	Turkiye Radyo-Televizyon Kumuru
United Kingdom	British Broadcasting Corporation
	Independent Broadcasting Authority
Vatican State	Radio Vaticana
West Germany	Arbeitsgemeinschaft der Öffentlichen Rundfunkanstalten
	Zweites Deutsches Fernsehen
Yugoslavia	Jugoslovenska Radiotelvizya

million television households, while the United States paid $225 million to reach 84 million households. By purchasing programming as a block, they eliminate competition, so sellers of international events must meet their terms or go without a sale to Europe entirely.[5]

The EBU is a non-profit organization, with an annual budget of about $7.5 million. The four major sources of funding are:

1. Active membership subscriptions
2. Associate membership contributions
3. Interest on EBU savings
4. Sale of publications and miscellaneous income

The annual contribution is based upon the size and estimated number of licence fees or sets within the country.[4]

Within the EBU, Eurovision is the most visible part of the organization's programme exchange. When the EBU was first established, most exchanges involved the physical transfer of film or tapes. Live transmissions were very difficult with the three line standards (405, 625, and 819), and were often of poor quality. In 1954, the Eurovision Control Centre opened in Brussels, and during the sixties, a more permanent network was established.

Eurovision currently reaches about 100 million receivers and over 300 million television viewers. The news exchange occurs in three daily conferences during which each member offers items and indicates whether it wishes to participate in the transmission of items from other countries. Three daily news broadcasts follow the conferences. In 1982, *Eurovision News Exchange* provided 7636 items.[6]

In 1982, 1942 hours and 936 programmes were exchanged within the EBU. Sporting events dominate the non-news programming, accounting for over 80% of the transmissions. While European sports are those most widely shown, recent attention has been focused upon other major outside events, such as the Olympics, World Cup Football, inaugurations, and elections. Other EBU programmes, such as *Jeux sans Frontiers* (*It's a Knockout*) and the *Eurovision Song Contest* are popular European-wide.

The EBU has managed to create some unity among the very diverse broadcasting systems within Europe, and has tackled some issues of pan-European television. It has served as an example for other broadcasting and programme exchange organizations on a worldwide scale.

2.3 Regulations in Europe

The close proximity of countries and overspill of television signals has caused difficulty in formulating national television regulations. As a result, a number of pan-European regulations have emerged. Not all were developed specifically for telecommunications.

2.3.1 TREATY OF ROME

In 1957, the EEC Convention ratified the *Treaty of Rome* which provides for the free flow of goods and services, unrestricted competition, and free market practices. A European Court of Justice ruling has defined broadcasting as a service to which the free trade rules apply.

Interpretations of the *Treaty of Rome* by the European Court of Justice have recognized that some advertising regulations were *de facto* created to discriminate against foreign products and should, therefore, be removed. They have also confirmed that any product freely available in one member country should be available in another.

The *Treaty of Rome* has several implications for broadcasting. First, government-operated broadcasting channels have been treated as a service. As a result, signal spillover has also been ignored, even when advertising practices and content rules vary. Second, programming distributed via satellite has been considered as a service, which would allow for authorized signal reception throughout the EEC. This would remove some of the strength held by the government-run broadcasting monopolies. Finally, pan-European broadcasting services are being encouraged, which alters the entire nationalistic orientation of the government broadcasters.[7]

2.3.2 EUROPEAN AGREEMENT ON THE PROTECTION OF TELEVISION BROADCASTS

In 1960, the 21 member countries of the Council of Europe instituted the European Agreement on the Protection of Television Broadcasts. It is a legally non-binding agreement that entitles states to authorize or prohibit CATV distribution in order to protect the existing governmental broadcasting channels. Some European countries are highly cabled, so the Agreement gave governments the option to increase or decrease protection of the monopolies held by their own channels.[8]

2.3.3 AGREEMENT FOR THE PREVENTION OF BROADCASTING TRANSMITTED FROM STATIONS OUTSIDE NATIONAL TERRITORIES

In 1965, the Council of Europe instituted an agreement to protect state broadcasters. Entitled Agreement for the Prevention of Broadcasting Transmitted from Stations outside National Territories, it is better known as the Pirate Broadcasting Law. The agreement prevents reception of television signals by signatories from channels unauthorized by another country.

At the time, commercial radio and television channels were being broadcast from ships on international waters which were taking a significant share of viewers and listeners from the government-run, non-commercial stations. Under the Agreement, unlawful activities include provision of equipment, ordering of production, ordering of advertising production, and provision of related services for a pirate broadcasting station.[9]

This regulation has recently received considerable attention because the ruling can also be applied to broadcast stations on aircraft or airborne objects, including satellites. As a result, delays in programme distribution via satellite to cable have arisen because some countries have not yet determined how to regulate the satellite systems. For example, Sky Channel, a UK-originated satellite-delivered television channel, was not cleared for reception in the United Kingdom for some time. It was, therefore, unlawful for the signatories to the Pirate Broadcasting Law to receive Sky Channel.

2.3.4 ADVERTISING IN RADIO AND TELEVISION

In 1981, a legally non-binding agreement was reached by European countries on advertising in radio and television. It suggested that the increased border flow of television and new media technologies had merely increased an ongoing problem of regulatory differences and recommended that the countries reexamine existing restrictions on the advertising of certain products and services with a view to harmonization. It recommended that a greater effort be made to achieve unified practices in broadcasting which would lead to fewer problems over programming content as the new technologies emerge.[8]

2.4 Revenue

Revenue sources for the government-run television stations are derived from a combination of licence fees and advertising. In nearly every country,

Table 2.2 Estimated broadcast revenue*, 1983[10–12]

Country	Licence fees ($ million)	Advertising revenue ($ million)	Total revenue ($ million)
Austria	245	135	380
Belgium	238	44	282
Denmark	195	0	195
Finland	139	75	214
France	889	678	1 567
Germany	1796	727	2 523
Greece	0	79	79
Ireland	43	48	91
Italy	0	817	817
Netherlands	244	99	343
Norway	119	0	119
Portugal	32	31	63
Spain	0	467	467
Sweden	256	0	256
Switzerland	154	63	217
United Kingdom	1134	1588	2 722
Europe	5484	4851	10 335

* Excludes government subsidies to broadcasters.

Table 2.3 Viewing levels of European television channels[10, 13]

Country	Average television viewed per day	Channel	Average reach (%)	CPT ($)
Austria	2h 18min	ORF1 and 2	26	5.0
Belgium	2h 55min	RTL	8	5.3
Denmark	1h 54min	DR		
Finland	2h	MTV1 and 2	35	7.1
France	2h 9min	TF1	6	9.7
		A2	8	5.9
		FR3	4	9.6
Greece	3h	ERT1 and 2	14	2.9
Ireland	2h 19min	RTE1	23	1.9
		RTE2	10	1.4
Italy	2h 40min	RAI1 and 2	17	1.9
		Canale 5	13	2.3
		Rete 4	7	2.5
		Italia I	8	2.5
		Euro-TV	4	2.8
Netherlands	1h 27min	NOS 1	7	7.6
		NOS 2	3	12.4
Norway	1h 18min	NRK		
Portugal	3h	RTP1	15	1.7
		RTP2	4	3.9
Spain	3h	TVE1	46	1.9
		TVE2	19	2.2
Sweden	2h	STV1		
		STV2		
Switzerland	1h 18min	DRS	4	22.2
		SR	1	26.2
		SI	1	13.9
		TOTAL SSR	6	16.9
United Kingdom	3h 10min	ITV	18	9.9
West Germany	2h 13min	ARD	19	3.2
		ZDF	17	2.5

a tariff is levied upon a television receiver set. The revenue is used to produce programming and fund the organization. In some countries, such as Denmark, Sweden, and Belgium, the licence fees are the sole source of funding. Revenues are listed in Table 2.2 and a summary of average viewing time, reach, and cost per thousand figures for the channels are outlined in Table 2.3 and then discussed in detail for each country.

Most television advertising time must be purchased well in advance, and this period can be over a year prior to the air date in countries such as Austria, the Netherlands, and West Germany. There is very little flexibility as to when a specific advertisement will be aired, yet demand generally far exceeds supply for airtime. The limited advertising time that is available is oversubscribed in many countries (see Chapter 3).

2.5 Country profiles

The following section details the television and, to a lesser extent, the radio broadcasting industries within each country, with respect to regulatory organizations, programme content, advertising and revenues, and future developments.[14]

2.5.1 AUSTRIA

The broadcasting authority in Austria is the Österreichischer Rundfunk Fernsehen (ORF). The ORF administers television and radio programming on FS1 and FS2 for television and Austria Regional, Austria Three, and Austria One for radio.

About 75 hours of programming are broadcast weekly on both FS1 and FS2. The average Austrian viewer watches 2 hours 18 minutes of television each evening. The largest programme categories are cinema and films (15.7%), made-for-television films (12.7%), and news and current affairs (12.4%). Entertainment programming accounts for over one-third of all of ORF's programming.

The Austrians are relatively positive about their broadcast services, and television has an average peak rating of 26% per evening. Foreign programming is available, primarily from Germany and Switzerland, and it receives a 7% share of evening viewing, most of which is after 20.00 hours, by which time the Austrian advertising blocks are over. Over 30% of Austrian viewers watch foreign programming at some time.

Of funds received for the television and radio channels 30 to 40% are obtained from licence fees and the remaining share from radio and television advertising. Licence fees of 1840 Austrian schillings annually contribute $245 million to operating costs of the ORF. Advertising, contributed $94 million towards operating costs in 1983.

The ORF is responsible for the sale of advertising time. Advertising is sold in September of the previous year and is broadcast daily except Sundays. Advertising is broadcast in blocks of 4 to 5 minutes between programmes. A maximum of 20 minutes per day of advertising is permitted. The average CPT for peak-time viewing is $5.00.

Commercial airtime on television is sufficient for demand, so Austria sees no reason to extend its commercial broadcasting. In addition, the Austrians want to keep the plurality of opinion within their newspapers flourishing— another reason not to expand commercial television.

With respect to the three national radio stations, Austria Regional (AR) and Austria Three (O3) are commercial, but, as with television, Austria

One (O1) carries no advertising. The average reach of AR and O3 is 29 and 22% respectively. The total daily radio reach is 78%.

Total commercial time on radio is limited to 120 minutes per day for the combined channels. Commercials are transmitted in blocks. Radio advertising contributes over $40 million to operational costs of the ORF.

The Austrians are currently involved in the German language 3-SAT project. The ORF is contributing programmes from FS1 and FS2 to the general entertainment channel to be broadcast via satellite to cable systems (see Chapter 8).

2.5.2 BELGIUM

The primary legislative act for broadcasting in Belgium was established under the 1960 Organic Law. It divided the broadcasting industry into two language groups, with full organizations for each: French—Radiodiffusion Télévision Belge (RTB)—and Flemish—Belgische Radio en Télévisie (BRT). Each organization operates two national channels.

Colour television was launched in January 1971, and during 1974 to 1975, television services were restructured. A third broadcasting entity was established in 1977, the Belgischer Rundfunk. It is a German language channel targeted primarily to the area of Belgium closest to Germany.

Each board of directors (for BRT and RTB) is appointed by the Dutch Cultural Community Council and the French Cultural Community Council. A Director General runs the daily operation in each organization. The RTB and BRT draw up individual programme plans which are then approved by their respective board of directors.

Belgium is the most highly television-saturated country in Europe, with access to nine foreign stations (three French, three German, two Dutch, and one Luxembourg) in addition to its own four national channels.

Together, the BRT and RTB broadcast over 3000 hours of programming annually. Over 70% is foreign programming, dubbed into French and subtitled into Dutch. The programme categories by type broadcast during the schedule is as follows:[15]

Fiction	24%
Documentary	3%
Information	12%
Magazine	6%
Variety	8%
Sports	9%
Game/quiz	3%
Theatre/music, classic	10%
Others	25%

Viewers' reactions to the Belgian broadcasting service are mixed. Most viewers would like to see even more choice in channel availability. They react against the government monopoly and political trend of services, but this is offset by the proliferation of choice through foreign programming and cable television. Viewing of foreign programmes is high, but generally limited to those who speak the respective languages. The average viewer watches close to 3 hours per day, one of the highest viewing levels in Europe.

The funding for the channels is entirely through government subsidies although a contribution to revenue occurs through an annual licence fee for each television household. The annual licence fee is 4500 Belgian francs for a colour television set. Over $238 million is contributed towards operating costs annually by the licence fees.

A 1910 regulation, reaffirmed under the Law of 18 May 1960, forbids advertising on the airwaves, and Belgium is one of the few countries with no advertising at all on its state-run channels. However, Belgium is one of the most cabled countries in Europe, and commercial television is available via the cable systems. In addition, Radio Télé-Luxembourg (RTL) is directed in part towards the French section of Belgium, and advertisers can make use of its commercial time. On RTL, advertising is broadcast daily, with 6 minutes per hour and a maximum of 68 minutes per day. Advertising is broadcast during 20 blocks. The peak-time CPT for TRL is $5.30, at which time the average peak rating is 8%. Sales of advertising time are preferably 3 months in advance.

The BRT and RBT each broadcast three radio channels each: BRT1, 2, and 3 and RBT1, 2, and 3. They are non-commercial stations funded entirely through licence fees. There is also a seventh channel broadcast in German. The BRT channels have a daily reach of 50.7% while the RTB channels have a daily reach of 37.1%. Local radio channels have a daily reach of 46.6%. In addition, Belgium receives several other foreign radio stations that are quite popular: Europe, RTL, and France Inter in French, Hilversum 1 and 2 in Dutch, and Radio 1, 2, 3, and 4 from the BBC.

As of 1979, more flexible arrangements were introduced for the authorization of local radio stations. Stations must be recognized by their local cultural council in order to broadcast and they must promote civic values. The stations are run on a non-profit and non-commercial basis. The authorization for local stations to operate is granted for 2 years.

The Belgian government has decided to permit some advertising on radio and television to be introduced in late 1985 or early 1986. A newly formed committee has been established to monitor advertising. It will be permitted in blocks only, and political, philosophical, religious, and ideological advertisements as well as advertisements for trade unions are forbidden. Ad-

23

vertising is not permitted before or after programmes for children under 16 years of age.

Part of the television revenue will be given to the print media and the remainder to RTB and BRT. Radio stations will be permitted to keep all of their advertising revenues. The BRT is considering a commercial television network in the Flemish area, in the event the advertising ban is changed. A group of Belgian newspaper owners are the primary investors who wish to expand their media interests.

The French RTB is a contributor to TV-5, the French language general entertainment service offered to cable systems via satellite (Chapter 8). It is interested in the possibility of establishing a pay-television channel in the French area, and is considering establishing a group of local television stations run by non-profit organizations and devoted to education.

2.5.3 DENMARK

Radio Danmarks has a government monopoly of both television and radio broadcasting under the 1971 Danish Broadcasting Act. The internal structure of Radio Danmarks is similar to that of the BBC. It is a public corporation and is governed by the Radio Council, with 24 members, 22 appointed by the parliament and 2 by the staff of Radio Danmarks.

Programme transmission of Denmark's single television channel is 6 hours 30 minutes each evening. Educational and children's programmes are broadcast during the day. Programmes in the evening are also focused upon education and information programming. The average viewing time for television each evening by adults is 1 hour 54 minutes.

There is general dissatisfaction by Danish viewers with the Danish service: 50% of viewers receive the three German channels and 10% receive both Swedish and German, and 37% of viewers watch foreign programming regularly, with over 63% of viewers watching it at some time. Of the 30% who are able to receive Swedish television, 12% of viewers watch it every day.

The single television channel is funded entirely through licence fees, as commercial broadcasting is forbidden by law. Licence fees are 1080 Danish kroners annually for a colour television, giving Radio Danmarks an operating budget of $195 million.

Radio Danmarks operates three radio channels that are also non-commercial and focused upon educational programming. In addition, in 1982, a law was passed to begin local radio transmission and experiments through regional programming.

The Danish constitution calls for absolute freedom, and the distribution of television signals by satellite has forced the government to confront the monopoly broadcast issue. It is apparent that both the government and Danish people believe that the time for a change has come and their broadcast system will reflect these changes with a new channel and the possible acceptance of commercial television in some form.

Over 75% of the Danish citizens would like an additional television channel, either commercial or non-commercial. Legislation to establish a second land-based channel was introduced by the Danish government in the autumn of 1984. While the government and the majority of viewers are in favour of the introduction of advertising, there is still some sentiment against it. The government intends to launch the channel without advertising if a majority does not vote in favour of it. They are also interested in establishing local television programmes without advertising, except for non-commercial interests, such as community organizations.

2.5.4 FINLAND

Broadcasting in Finland is operated by the wholly state-owned corporation Oy Yleisradio Ab (YLE). The Council of State is selected by Parliament every 4 years and it decides upon YLE's operations to ensure that various political parties and interest groups have equal opportunities to express their opinions. The Minister of Communications oversees the operations of YLE on behalf of the Council of State. YLE is owned 99.9% by the state, with the remaining 0.1% divided among 55 shareholders.

YLE operates both of Finland's national television channels and is non-commercial. It sells transmission time to Oy Mainos-TV-Reklam Ab (MTV), an independent Finnish commercial television company that finances its operations by selling advertising time. MTV buys 19 hours 20 minutes of transmission time from YLE weekly: 12 hours 45 minutes on YLE1 and 6 hours 35 minutes on YLE2.

The categories of programme type transmitted by YLE are as follows:[17]

Time devoted to:	Broadcasting (%)	Viewing (%)
News	12	15
Current events	5	6
Documentaries	16	8
Educational	7	1
Theatre	2	2
Film: full-length and television	15	20
Series	12	19
Music/light entertainment	12	15
Other	19	14

Nearly 60% of overall programming on both channels is foreign. Weekly transmission time per channel is approximately 30 hours. YLE/MTV1 and

2 broadcast 4438 hours annually. Of these, Finnish regional and local programmes comprise 338 hours, Swedish comprises 998 hours, and Lappish comprises 128 hours.

Finland is the only member of both the EBU and OIRT broadcasting organizations. It is also a member of Nordvision.

In general, the Finns are very positive about their television services. Close to 85% of the population views television at least once per day, average peak rating is 35%, and average daily viewing is 2 hours. This rises to 3 hours on weekends.

The YLE owns and operates the network of land transmitters for the YLE/MTV1 and 2. Licence fees are 590 Finnish kroners for YLE1 and 2, which contributes $139 million towards operating costs annually. Advertising revenues through MTV for 1983 were $75 million.

MTV sells up to 15% of its airtime to advertisers—about 9 minutes per hour. Advertising is sold three times per year, about 2 to 3 months before each triannual programme period. Advertisers can specify the date and time of transmission, so targeting of audiences is possible. The Finnish television channels are the only commercial ones in the Scandinavian countries. Advertisers must follow MTV's Principles of Advertising which adheres to the International Code of Advertising Practice.

Oy Yleisradio Ab, the Finnish Radio Corporation, operates three channels which are non-commercial. Group A covers serious music, educational programmes, current affairs, talk programmes, radio dramas, light entertainment, religious programmes, and children's programmes. Group B covers light music, programmes for youth, talk programmes, sport, and serious music. Group C is a Swedish programme.

Finland has been very progressive with its media services, and its first cable operator, HTV, was established well over 10 years ago. The country is interested in further cabling, and the broadcasting industry is involved in Tele-X, the DBS Scandinavian satellite consortium (see Chapter 5).

2.5.5   FRANCE

French broadcasting is held by the state as a monopoly under the Secretary of State for Posts. It is operated through Télédiffusion de France (TDF) under direction from the Minister of Communications. On 29 July 1982, Law No. 82.652 was enacted concerning broadcasting and communications. With the introduction of new media technologies, the government monopoly on broadcasting was difficult to uphold, and the new law allows a general right to public audio-visual information. In addition, programming was decentralized to free it from political influence; this independence is guaranteed by the creation of the Haute Autorité for Audio Visual Communications. Four programme contractors now work under TDF: one for radio—Radio France—and three for television—TF1 and Antenne 2 (A2), the national channels, and FR3, a showcase channel for the 12 regional television production centres.

The Haute Autorité has nine members appointed for nine years by the President. The Autorité has authority over the programming of the three television channels. There is no national restriction on foreign reception of national broadcast stations and French viewers are free to tune into foreign stations.

TF1 and A2 transmit approximately 83 hours of programming per week, while FR3, the regional station, transmits some 63 hours per week. The average peak rating for adults is: TF1—6%, A2—8%, and FR3-4%. In addition, Radio Télé-Luxembourg and Télé-Monte Carlo reach 8.7 and 5.8% of French viewers respectively. The French have one of the lowest opinions of their national television. They are noted for their poor overall use of television and even the Haute Autorité has issued a report that states French television is overbureaucratic, has a low productivity level, and an inability to absorb technological change.

Despite this fact, France is currently the fastest growing market in Europe for television viewing and radio listening. An average viewer watches television 2 hours 9 minutes per day.

Of the programming on the French channels 13% is foreign. The largest programme categories are fiction (21%) and magazine (18%). The French would like to see more news, better teenage programmes, fewer quiz shows, and better films.[15]

Funding is through a combination of licence fees and advertising. Governmental guidelines suggest that advertising supplies no more than 25% of the funds; however, with the recent addition of advertising on FR3, it now accounts for 33% of total revenue. Licence fees are 471 French francs for a colour television set, contributing $889.47 million in licence fees annually. Advertising revenues for the three channels contributed $452 million towards operating costs in 1983.

Content rules on advertising are supervised by the Haute Autorité and Régie Française de Publicité. Advertising is broadcast on all three channels. A daily average of 18 minutes of commercials is run during the evening hours, and advertisers may specify a specific date and time for transmission. Commercial breaks are not permitted to interrupt programmes. The average peak-time CPT for all adults is about $9.6 for TF1, $5.9 for A2, and $9.6 for FR3. While advertising demand exceeds supply for advertising on the state-run channels, Radio Télé-Luxembourg in Luxembourg and Télé-Monte Carlo in Monaco offer commercial time in the French language directed at viewers in France.

In 1983, the decentralization process reorganized the radio industry, permitting regional and local radio companies for public service broadcasting. In addition to the regional radio stations, France operates three national radio stations under the direction of Radio France: France-Musique, France-Inter, and France-Culture. There are also three commercial radio stations available in France: Radio Télé-Luxembourg (RTL), Radio Monte Carlo and Europe No. 1. While RTL and RMC are broadcast outside France, SOFIRAD, a state-run organization, has minority ownership of Europe No. 1. Radio advertising revenue generated in France in 1983 exceeded $7.5 million. The regional radio companies operate local radio stations, and the government has allocated over 700 private licences to operators. Local radio stations are forbidden to carry advertising, being funded out of local government revenues.

French broadcasting is developing in several areas. Canal-Plus, a fourth terrestrial channel, began broadcasting in November 1984. It is a pay-television service, broadcasting films and some other general entertainment. Its signal is scrambled, except for a small period of time each day when it carries advertising (see Chapter 8).

All three national broadcast channels are involved in TV-5, the French language general entertainment programme distributed via satellite to cable service. TF1, A2, and FR3 supply five-sevenths of the programming, contributing their best productions. In addition, TDF is also interested in DBS programming and plans to provide services via the TDF DBS satellite in 1987 (see Chapter 5).

In January 1985, President Mitterand announced that he would permit private television broadcasting in France. He suggested that approximately 80 local channels could be connected into two national advertiser supported networks. He commissioned a report by a French lawyer Bredin, who reported in May 1985 essentially supporting the Mitterand proposal. In August 1985 the Government moved closer to implementation of its new policy by endorsing the two new network concepts and suggesting that one would be mainly music and the other mainly entertainment and information programmes. The nationwide networks would be able to reach about 40% of the French population and would be supplemented by 40 to 50 new local television stations. The networks will have expanded opportunities for advertising. There being a prohibition on commercials which interrupt programmes, and the local television stations will be subject to strict media cross-ownership rules. No owner of a regional paper could own a local TV station in that region.

Network programmers were not announced at the time. In December 1985, the first new channel was allocated to a consortium of French industrialists (Jerome Seydoux with 40% and Christophe Riboud with 20% and the Italian entrepreneur Silvio Berlusconi with 40%). An entertainment channel, it is advertiser supported and will also take a transponder on TDF (see Chapter 5). Robert Hersant, one of France's largest press groups, has also announced more detailed plans to establish private television services.

It has announced the creation of TVE (Teleurop) which plans to broadcast 18 hours per day and is also a contender for a national channel (A2) in the event that French television is privatized.

2.5.6 GREECE

The broadcasting industry in Greece is controlled by the Ministry to the Prime Minister. Elliniki Radiophonia Tileorassis has operated two national channels ERT1 and ERT2 since 1968. They cover about 90% of the total population in Greece. Weekly transmission times are between 58 and 66 hours per channel. Over 40% of the programming on the two channels is foreign with subtitles. There are many government-controlled political broadcasts, and dislike for these programmes is high. Despite a low public opinion of the broadcasting system, the Greeks have a high level of television consumption, exceeding 3 hours daily. The average peak rating is about 40% for both channels.

Where foreign television is available, many viewers watch it despite the language difference. In western Greece, viewers prefer to watch Italian television, while those in northern Greece watch Yugoslavian television.

ERT1 and ERT2 compete for audience and advertising revenue. ERT1 is funded through a combination of advertising and government funds, while ERT2 is completely funded through advertising. There is no viewer licence fee. Revenue generated from advertising reached $56.70 million in 1983. The average CPT during peak viewing time is $2.9.

Advertising is broadcast daily. ERT1 broadcasts commercials in four 5- to 10-minute blocks, usually before the most popular programmes. ERT2 broadcasts commercials throughout its transmission. Commercial time on ERT1 is limited to 30 minutes per day, while ERT2 transmits 35 to 45 minutes per day. Advertising time is sold on a monthly basis in the first 2 weeks of the previous month.

The Greeks have no immediate plans to alter the present system or move into other areas of new media.

2.5.7 IRELAND

Television broadcasting in Ireland is operated by Radio Telefis Eireann (RTE), a semi-state organization. The RTE was established under the Broadcasting Act of 1960 and is governed by the RTE Authority, a nine-member board appointed by the government for a period of up to 5 years. There are nine major divisions, and the Director General runs the daily operation in a structure similar to the BBC. TV1 began in 1961, while TV2 was established in 1978 to provide further viewing choice and programming for minority audiences. The Broadcasting Authority is responsible for regulating the RTE and overseeing RTE's obligations to serve the public interest, uphold democratic values, and report news objectively and impartially.

RTE1 is the main channel for news, current affairs, sport, and general interest programming. It is an active member of Eurovision, and uses its news feeds three times daily. RTE2 is primarily a minority interest channel. The average peak rating for RTE1 is 23%, and for RTE2, 10%. During prime time, viewing of RTE1 can reach as high as 50%. As 20% of the population speak Irish, a portion of their programming is devoted to this audience. The average viewer in Ireland watches 2 hours 19 minutes per day.

Over 55% of the programming broadcast is foreign, supplied primarily from the United Kingdom and the United States. The programmes that the RTE does produce are noted for their high quality because they must compete with the BBC and ITV for audiences. Nearly 50% of Irish viewers can receive British television which receives about a 25% share, and close to a 40% share on the east coast. In general, the Irish are very satisfied with their programming.

The two television channels, RTE1 and RTE2, and the radio channels are funded by licence fees (41%), television advertising (37%), and radio advertising (13%). The remainder comes from contributions to RTE. Licence fees are £52.00 annually for a colour television set, and contribute $43 million towards the operating budget of RTE. Advertising revenues contribute $33 million annually.

The 1960 Broadcasting Act and 1976 Amendment regulate commercial broadcasting. The total daily commercial time is fixed by the Broadcast Authority and confirmed by the Minister of Posts and Telegraphs. In addition, the RTE has a code to control commercial advertising.

Both television channels are commercial. Advertising is limited to 10% of transmission time daily, with an average of 7 minutes per hour. Commercials are transmitted both between and during programmes. Advertisers can specify the data and time of commercial broadcast, and the top-rated programmes are sold at the most expensive rates. The average CPT is $1.9 for RTE1 and $1.4 for RTE2. Advertising time is sold on a first-come first-served basis 1 to 3 months prior to transmission time.

The RTE operates two commercial radio stations, Radio 1 and Radio 2. Radio 1's peak-time reach is 22 to 37% for adults, while Radio 2's audience is predominantly the youth market. Radio Na Gaeltachta is also a national service which broadcasts for the Irish-speaking communities.

2.5.8   ITALY

Italy is in a unique situation because it has a combination of state-operated and private television. Hundreds of local television channels began opera-

30

tion after two decisions by the Constitutional Court in 1974 and 1976 removed Radiotelevisione Italiana's (RAI) monopoly over local broadcasting.

RAI has operated as a public organization since 1924. It is 99.55% owned by the Istituto per la Ricostruzione Industriale (IRI) and 0.45% by the Societa Italiana Autorite Editori (SIAE). RAI is granted a licence by law under agreements between the Ministry of Posts and Telecommunications and RAI. The President of RAI is the Chairman of the board of directors which supervises RAI's management. The board of directors is composed of 16 members, 6 elected by the IRI and 10 elected by Parliament's Supervisory Committee. The Director General is responsible for the day-to-day operation. Both the Director General and the Chairman are appointed by the board of directors.

While RAI1, 2, and 3 operate under government control and are restricted to local programming only, private television is practically lawless. As a result, the private television stations have become popular, and through video cassette networks, pose a severe competitive threat to the government networks.

Despite the local television law, the private stations have managed to build several prominent networks and syndicate their programming through the use of video cassettes and prearranged advertising. A 1976 ruling by the Constitutional Court prevents the interconnection of the private stations, retaining that monopoly for RAI as the national broadcaster, so it is difficult for these stations to broadcast any news. As a result, the news on the RAI channels is about the single most popular programme. Of the 400 or so private television stations, about 100 to 120 are divided into four popular 'networks': Canale 5, Rete 4, Italia 1, all owned by Silvio Berlusconi, and Euro-TV. Berlusconi, also positioning his company to enter the DBS field, has created a network which rivals the national broadcaster in viewing size and exceeds it in advertising revenue. These four networks account for about 90% of the private television audience at prime time. About 100 stations remain stable with programming, the remainder often being erratic.

Weekly transmission time for RAI and the major private television stations are: RAI1 and RAI2—86 hours; RAI3—40 hours; Canale 5 and Italia 1—16.6 hours; Rete 4—15 hours; and Euro-TV—14 hours. Italy is Europe's richest market for television products, spending over $150 million annually for foreign programming: 14% of RAI's programmes are foreign, while over 80% of the programming on the private stations is foreign. The government has been attempting to impose some restrictions upon the amount of foreign programmes shown on the private channels. The private networks draw heavily from US programmes, providing entertainment channels of films, series, quiz shows, and entertainment spectaculars. Films

31

and series provide about 35% and 20% of the programme schedules respectively.

Some foreign programming is viewed, particularly programming from Monaco, Switzerland, France, Malta, and Yugoslavia. Télé-Monte Carlo (TMC) broadcasts to the north of Italy. TMC was purchased by the RAI in 1983. TV Capodistria in Yugoslavia began in 1971 and provides coverage in northeastern Italy. The average peak rating in Italy is:

Italian	RAI1 and RAI2	17%
	Canale 5	13%
	Rete 4	7%
	Italia 1	8%
	Euro-TV	4%
Foreign	TMC	0.2%
	Capodistria	0.1%

The Italians are known to be very critical of their television services. However, the expanded choices of private television have proved to be popular. Viewing is over 2 hours 40 minutes per day—one of the higher figures in Europe.

The RAI channels are funded through a mixture of licence fees and advertising, while the private stations are funded entirely by advertising. There is rarely enough advertising on the RAI channels to reach the 5% limit that is imposed. The RAI channels attempt to reach a 10% level during prime time when the rates are higher. Advertising on RAI is transmitted in blocks at fixed times, between and during programmes. There are no limits on advertising per hour, with 59 to 69 minutes broadcast daily for RAI1 and RAI2, and advertising revenues exceeded $654 million in 1983. The average CPT for RAI1 and RAI2 is $2.

Private television operations have agreed not to exceed 14% of total content for advertising, and about 9 minutes of advertising is broadcast per hour. However, this transmission of advertising often reaches 30% during prime time. Commercials are broadcast at regular 15-minute intervals. CPTs for private television networks are: Canale 5—$2.33; Rete 4—$2.46; and Italia 1—$2.46.

RAI operates two national radio stations, RAI1 and 2, and 20 regional stations under the umbrella of RAI3. All are commercial channels. Each RAI channel has an approximate reach of 20 to 25%. Radio in general has a daily reach of 60%, while local radio has a daily reach of 30%.

Private radio, like private television, is permitted, and over 2000 regional and local stations are operational. About 500 of these channels have good-quality programming and are commercial. Several of the major private stations are: Studio 105, Gamma Radio, Sper, Margherita, Tir, and

Cepe/Studio D. About 10 minutes per hour of advertising is broadcast, accounting for $163 million in advertising revenues in 1983.

The private television industry in Italy has precluded the need for any further expansion of broadcast services. However, RAI is interested in further distribution of their programming by satellite, and is involved with the Olympus DBS project and ECS.

2.5.9 NETHERLANDS

The Netherlands has a unique broadcasting situation that is regulated by the 1967 Broadcasting Act under which the Authority of the Ministry for Cultural Affairs, Recreation, and Social Welfare is responsible for the broadcasting industry. If specific groups of the population and ideological and religious groups meet certain conditions, the Act allows them access time to television. Initially, the categories of broadcasters are determined by size:

A	450 000 members or more
B	300 000–450 000 members
C	100 000–250 000 members

Access is also given to 'aspirant broadcasters' with 40 000 to 100 000 members.

These organizations are designed to ensure that broadcasting remains open, diverse, and cooperative, and that it creates a favourable climate for free access to broadcasting for the greatest range of groups and ideologies. There are eight major broadcasting organizations five category A, one category B and two category C.

The Dutch Broadcasting Foundation, the Netherlands Omroep Stichtine (NOS), acts as the umbrella organization for the Dutch broadcasters, providing facilities and working staff, as well as coordinating and producing programmes. NOS was founded in 1969 when the Dutch Radio Union (NRA) and Dutch Television Foundation (NTS) merged. Programme committees issue directives and guidelines on NOS programmes to the management board. The committee consists of members from both broadcasting organizations and organizations with a cultural, religious, or philosophical basis.

The NOS has a board of governors whose Chairman is appointed by the Minister for Cultural Affairs, Recreation, and Social Welfare. A Broadcasting Council advises the ministry on all questions concerning radio and television, and is overseen by a government Broadcasting Commission who is appointed by the Crown. Day-to-day operations are the responsibility of a management board whose members come from both broadcasting organizations and organizations with a cultural, religious, or philosophical basis.

Nederland 1 and Nederland 2, the two national public broadcasting channels operated by NOS, broadcast about 50 hours and 30 hours weekly, and average peak rating for both channels is: NOS1—7% and NOS2—3%. Programming by category is as follows:[15]

Fiction	21%
Documentary	17%
Information	9%
Magazine	15%
Variety	11%
Sport	8%
Theatre/music, classic	4%
Other	15%

Dutch viewers watch 1 hour 27 minutes of programming per evening. This is one of the lowest figures in Europe.

The majority of Dutch viewers are satisfied with their television service despite the fact that Dutch television is often considered to be the worst in Europe. In fact 63% of viewers are satisfied with the choice of television programmes and 52% are satisfied with the quality of programmes. The young are the most critical, and their complaints include: lack of films, plays, nature documentaries, entertainment, and sport and too many quiz, discussion, and political programmes. Adults complain of a lack of news and sport programming. Of the programming broadcast 33% is foreign.

The Netherlands have numerous foreign channels available to them: ZDF and ARD from Germany, BRT1 and 2 from Belgium, BBC1 and 2, ITV, and Channel 4 from the United Kingdom, and some French television. Over 84% of the viewers can receive some form of foreign programming and over 60% can view a programme in a foreign language. During prime time, television sets are tuned to a foreign station 20% of the time.

Advertising provides 12% of the funding, the remainder being from licence fees. Licence fees are 158 Dutch florins annually and contribute $244 million towards the operating budget. Revenues from advertising reach $81 million. Advertising is sold exclusively by the state foundation Stichting Ether Reclame (STER). The Advertising Council (Reclameraad) has control of advertising, monitoring STER's pricing policy and content of advertising. Both Nederland 1 and 2 carry advertising from Monday to Saturday. Each channel broadcasts 15 minutes daily in these blocks which are transmitted before and after the news. Demand exceeds supply by 70%, so applications for airtime must be submitted in September for the following year. Only the month of transmission can be specified. The average CPT for NOS1 is $7.7 and for NOS2 is $12.4.

There are three national commercial radio channels operated by NOS—Hilversum 1, 2, and 3. STER is responsible for all of their advertising sales. Advertising is broadcast daily in blocks and reached $18.2 million in revenues in 1983.

A fourth regional radio channel without advertising is also available in addition to BRT1 and 2 and RTB1 and 2 from Belgium, WDR1, 2, and 3 from West Germany, Radio 1, 2, 3, and 4 from the BBC in the United Kingdom, and RTL are also received. The average daily reach of all radio channels is 70%.

The Netherlands broadcasting industry has been changing rapidly. The NOS has agreed to a third television channel for non-commercial, minority, cultural, educational, and religious programming. This is intended to allow Nederlands 1 and 2 to focus more on entertainment programming. It will be funded by a slight increase in the licence fees (see Chapter 8).

2.5.10 NORWAY
The Norsk Riksringkating (NRK) is a state-owned and operated broadcasting organization. It offers one non-commercial television channel.

NRK broadcasts 38.5 hours per week. It is gradually expanding its weekly output and plans to pass 60 hours by 1990. About 55% of the programming on NRK is original and is in Norwegian, while about 40% of the programming broadcast is foreign. About 50% of its foreign programming is in English, with an equal share of US and UK products.

Norway has the lowest average daily viewing by adults, and the NRK is often considered to be too boring and educational. Many viewers criticize the NRK as being too left wing in politics and culture, and a recent survey indicates that 62% of viewers would like a wider selection of channels; 33% of the viewers receive Swedish channels, and others receive British, Danish, and Finnish programming.

Funding of the channel is entirely through licence fees; 725 Norwegian kroners for a colour set gives the NRK an operating budget of over $119 million annually. While there is no advertising, there is a definite political leaning towards the introduction of a limited amount of commercial time.

The NRK operates two radio programmes, plus a regional service. All services are non-commercial.

The NRK has proposed that pay-television be introduced. It has proposed a model which would allow pay-television to reach 30% of the population in one year. However, the Conservative Government favours the introduction of a second television channel with limited amounts of advertising. Much debate has been going on, and the broadcasting system is likely to change in the near future.

2.5.11 PORTUGAL
The Radiotelevisao Portuguesa SARL (RTP) is controlled by a state-owned company broadcasting two television channels, RTP1 and RTP2. RTP1

provides national coverage, but RTP2 is restricted to the Lisbon and Oporto areas, the coastal areas, and southern areas, covering about 60% of the population.

RTP1 transmits over 90 hours per week, while RTP2 only transmits about 20 hours per week. Viewers watch over 3 hours of television daily, with an average peak rating of 15% for RTP1 and 4% for RTP2.

Viewers are not satisfied with the quality of programming. The introduction of colour television and the establishment of a new private company to produce programming are recent steps to improve television services. The most popular programmes are imported, and about 25% of the population watches Spanish television.

The two channels, RTP1 and 2, are funded by a combination of advertising and licence fees. Licence fees are 2500 ESC per year for a colour television set and 1250 ESC for a black and white set, contributing $32 million towards operating costs of the RTP. Advertising is broadcast daily throughout transmission in blocks between programmes and during natural breaks. There is a limit of 9 minutes per hour.

Advertising is sold by a separate company—Radio Televisao Commercial—and the supply of commercial airtime is adequate to meet the demand, except during November and December. The average CPTs for RTP1 and RTP2 are $1.70 and $3.90 respectively. Advertising revenues from television exceeded $25 million in 1983.

Two national radio broadcasting organizations operate in Portugal. Radiodifusao Portuguesa (RDP) is operated by the state and transmits RDP3 and RDP4. The other radio organization, Radio Renascenca (RN), is privately owned and linked closely to the Catholic Church. RDP3 and 4 are funded through a combination of advertising and licence fees, while RN is funded entirely from advertising. There are also two local radio stations.

Daily reach for each channel is as follows: RDP3—7.4%; RDP4—16.4%; RN—17.3%. Radio revenues are over $11 million annually.

The government has given approval for a new commercial channel owned by the Catholic Church. The launch date for this new service has not yet been determined.

2.5.12 SPAIN
The Spanish broadcasting industry is run by the state-owned organization Television Española (TVE). There are currently two national channels, TVE1 on VHF and TVE2 on UHF, and a regional network of four channels: two for the Canary Islands and one each for the Basque Country and Cataluna.

Weekly broadcasts on TVE1 and TVE2 are 60 and 35 hours per week respectively. Viewers in Spain are not particularly satisfied with their broadcasting service, and feel the general quality of programming is low. There is rising public discontent over the content and respective shares of political, religious, and entertainment programming. The average peak rating for TVE1 is 46% and for TVE2, 19%.

TV3, the regional television channel for Cataluna, began broadcasting in January 1984, transmitting 50 hours per week in Catalan dialect. Euskal Televista (ETB), the regional television channel for the Basque area, began transmission in January 1983 and broadcasts 30 hours per week in the Basque dialect of Euskera. The progamming on the two Canary Islands channels is identical with the two national stations.

About 20% of the programming broadcast on the state-run channels is foreign, and 30% of viewers watch foreign programming. The average daily viewing time for an adult is over 3 hours, one of the highest viewing levels in Europe.

TVE1 has carried advertising since 1969 and TVE2, since 1971. TVE is financed by a combination of advertising (90%) and a state budget (10%). Advertising revenues contributed $323 million to TVE's operating budget in 1983. There is no licence fee.

Advertising sales are administered by Gerencia de Publicidad de Television Española. Commercial time is broadcast daily with an average of 4 minutes per hour. Advertising is broadcast in blocks of not more than 12 commercials per block. The maximum minutes per day allowed are 48 during the week and 60 on the weekends for TVE1 and 42 for TVE2. Commercial time is sold on a first-come first-served basis, and if the requested block is full, an alternative is offered. TVE1 requires 4 to 5 months' advance booking and TVE2 requires up to 6 months' advance booking. The average CPTs for TVE1 and TVE2 are $1.9 and $2.2 respectively.

Spain has over 350 commercial radio stations, many of which are formed into networks. Radio Cadena Española (RCE) is state owned. It is commercial but also receives funding from licence fees. Sociedad Española de Radio difusion (SER) is privately owned and is funded entirely through advertising. Cadena Ondas Populares (COPE) is owned by the Catholic Church and is funded through a combination of advertising and licence fees. National reach for each channel is: RCE—4.6%; COPE—5.3; and SER—30.6%. Radio advertising revenues reached over $143 million in 1983. Radio Nacional de España operates three national programmes that are non-commercial.

Spain has recently decided to introduce private television in 1986. It plans to establish a private network composed of various private media and political opposition groups. This would give Spain four networks, with the 2 TVE channels, the regional network, and the private network.

2.5.13 SWEDEN
Both Swedish Television (STV) channels, STV1 and STV2, are operated by the government-run Swedish Broadcasting Corporation, Sveriges Radio.

The Swedish Telecommunications Administration has responsibility for the distribution of programmes produced by STV, and is also responsible for the collection of licence fees.

STV1 and STV2 transmit 54 and 43.5 hours per week respectively. Swedish viewers believe that STV1 and 2 are too culture oriented, and would like more entertaining programmes. Research indicates that 75% of the viewers are dissatisfied with the present quality of television broadcasts.

About 45% of the programming is foreign and many television programmes are subtitled for the benefit of the deaf. The average daily viewing time per adult is 2 hours. There is some reception of Danish, Finnish, and Norwegian television in addition to the Swedish offerings.

Funding is entirely through licence fees. Individual licence fees are 708 Swedish kroners for a colour TV and revenues generated exceed $256 million.

At present, there is no commercial television in Sweden. However, there is a continuous movement to allow advertising. Over 75% of the viewers favour advertising if the additional income goes towards programming. The Socialist Government is opposed to advertising, and while it remains in power there is no possibility that the regulations will change. Financial problems for the broadcasters are increasing and the Conservatives and Liberals are in favour of commercial television.

Sveriges Radio AB operates three internal non-commercial services. SR1 is news and talk programming; SR3 is light entertainment programming; SR2 includes minority programming. The channels are funded entirely by licence fees which are included with the television set.

The current agreement for Sveriges Television terminates in July 1986 and it is attempting to embark on a 10-year plan to move into new media areas. STV has plans for a third terrestrial network as a pay-television service. It is considering distribution to MATV systems, and may later use Tele-X or Nordsat, the Scandinavian satellites. From October 1984 until October 1985, STV have been experimenting with a pay service that offers films, sport, and entertainment for a monthly fee. An eventual acceptance of commercial television would also evoke considerable changes in the system.

2.5.14 SWITZERLAND

The Swiss PTT has the exclusive right to grant a concession to operate a service which transmits and receives signals. It has granted a monopoly for public radio and television to the Schweizerische Radio und Fernsehgesellschaft or SSR (the Swiss Broadcasting Corporation). The SSR is composed of three member groups which are each responsible for programming in a specific language: the Société de Radiodiffusion de la Télévision de la Suisse

Romande (French—24%), Radio und Fernsehgesellschaft der Deutschen und der Ratormanischen (German—72%), and the Societa-Cooperativa per la Radio-Televisionenella Swizzera Italiana (Italian—4%).

Weekly transmissions for the SSR are: German service—64.5 hours; French service—82 hours; and Italian service—52 hours. Average peak ratings across the networks are: German—3.7%; Italian—1.1%; and French—1.1%.

The largest amount of programming is cultural, over 40%, followed by news and information. Programming by type is as follows:[18]

Social and political programmes	8.3%
Cultural programmes	40.0%
Sport and leisure programmes	18.3%
Light entertainment	8.6%
News and information programmes	24.8%

Swiss viewers are not satisfied with their programmes and close to 40% of viewers watch foreign programmes during peak times. Every Swiss home can receive at least two and up to ten foreign television channels, and a large portion of the Swiss population can speak or understand other major languages. Programming from the French, German, Austrian, and Italian channels is popular.

Television is financed through a combination of advertising (40%) and licence fees (60%). Licence fees are 174 Swiss francs annually, contributing $154 million towards operating costs of the SSR. Advertising revenues contributed $61 million in 1983. Advertising is broadcast from Monday to Saturday in five blocks of 4 minutes on each channel, for a maximum of 20 minutes per channel daily. Commercial airtime is sold by a separate organization, AG fuer das Werbefernsehen, in the middle of October for the following year. A combined CPT for the total SSR is about $16.9, the highest in Europe.

There is no commercial radio in Switzerland, so the radio licence fees of 87 Swiss francs fund the stations. The German and French networks operate three channels, while the Italian network operates two. Südwestfunk 1 from Germany, Europe No. 1, and Radio Monte Carlo are all available in Switzerland and sell airtime to Swiss advertisers. As a result, about $2 million is generated in radio advertising revenues annually.

The SSR is involved with TV-5 in producing one evening's programming in French per week. It also has a minority interest share in Teleclub, a pay-television channel that offers films and entertainment to Swiss homes that are cabled.

2.5.15   UNITED KINGDOM
Under the 1949 Wireless Telegraphy Act, a broadcasting authority is required to be licensed by the Home Office. Such licences have been granted

to the British Broadcasting Corporation (BBC) and in 1955, the Independent Broadcasting Authority (IBA).

The BBC has two national channels, BBC1 and BBC2, while the IBA offers both regional and national coverage through 16 programme companies (the Independent Television companies (ITV)) and Channel 4 television company. In addition, TV-AM Limited provides the national breakfast time service across the ITV network. The IBA does not produce programmes on its own, but appoints the regional organizations to do so (Table 2.4).

Table 2.4 Programme companies of the IBA[19]

Station	Area served	Population coverage (million)
Thames TV	London	10.4
London Weekend TV	London	10.4
Central Independent TV	Midlands	8.5
Granada TV	North West	6.5
Yorkshire TV	Yorkshire	5.7
Tyne Tees TV	North East	3.4
Scottish TV	Central Scotland	3.5
Grampian TV	North Scotland	1.1
Harlech TV	Wales and West	4.8
TV South	South/South East	4.6
Anglia TV	East	3.7
TV South West	South West	1.5
Channel TV	Channel Islands	0.1
Ulster TV	Northern Ireland	1.4
Border TV	England/Scotland Borders	0.6
TV-AM	Total United Kingdom	66.2

The BBC was the first broadcasting organization in Europe, the first television service being introduced in 1936. It is a public corporation established by Royal Charter under the control of a board of governors appointed by the Privy Council and a Director General who operates the day-to-day affairs.

The IBA was established by the Television Act of 1954 as the Independent Television Authority, and was renamed by the Sound Broadcasting Act in 1967. The ITV companies began transmission in 1955 and Channel 4 began its transmission in November 1982. The IBA is appointed by the Home Secretary and a Director General also runs the day-to-day operations. The programme companies are granted a 12-year monopoly franchise in their respective regions and are subject to IBA policy and licence reviews.

The IBA members provide programmes in their respective regions to contribute to a national service. The Broadcasting Act of 1980 gave the authority to the IBA to provide a fourth television service, Channel 4, for minority programming in the United Kingdom. This national service is also composed of regional productions. The Broadcasting Act also provided for the establishment of a Welsh Fourth Channel Authority, which is responsible for a separate service on the fourth channel in Wales.

Both the BBC and IBA are required by law to set up special advisory committees to advise them on policy and programmes in general and on specialized areas. Both companies are required by law to 'educate, entertain and inform' the public, a model which has been favoured by many other government-operated broadcasting corporations in Europe. The IBA controls its programme planning and approves its members' schedules, while the BBC board of directors is responsible for ensuring compliance with its obligations. Under the 1980 Broadcasting Act, a new Broadcasting Complaints.Commission was formed to rule on unfair practices and complaints.

The United Kingdom broadcasts more hours than any other government-run broadcasting service in Europe. The hours per week broadcast by the channels are: BBC1—115 hours; BBC2—65 hours; ITV—127 hours; Channel 4—56 hours; TV-AM—17.5 hours. The average television viewer in the United Kingdom is 16 years of age and watches 3 hours 10 minutes of programming per evening, which is the highest daily viewing average in Europe. The programming and costs for programmes are as listed in Tables 2.5 and 2.6.

Table 2.5 Programming and costs for the BBC[20]

BBC	Output (%)	Cost (%)	Total cost per hour ($ thousand)
Drama	6	24	255
Light entertainment	10	16	108
Outside broadcasting events	1	1	97.5
Features, music, documentaries	13	18	84
Educational	3	5	79.5
Children's	8	7	54
Asian and community programmes	1	1	51
Religion	2	2	49.5
Current affairs	16	10	37.5
Outside broadcast of sport	27	13	30
Programme acquisition	13	4	18
Average cost per hour of all network first showings			73.5*

£1.00 = $1.50
* Figure not derivable from table as above includes repeats. Overall network average is $62 500 per hour.

Table 2.6   Programming and costs for ITV[20]

ITV	Programmes by type (%)	
News and news magazines	11.00	
Current affairs and general factual/arts	12.50	
Religion	2.25	
Adult education	1.75	
School programmes	5.50	
Pre-school education	1.75	
Children's informative	3.00	
INFORMATIVE TOTAL		37.75
Plays, drama, television movies	20.50	
Feature films	9.00	
NARRATIVE TOTAL		29.50
ENTERTAINMENT TOTAL		13.00
SPORT		19.75
Total		100.00

	Costs per hour ($ thousand)
Action/adventure	300
Drama series	157.5
Single plays, film	675
Single plays, tape	375–412.5
Prestige drama	675
Arts/performance/opera	240
Features educational	240
Current affairs (studio ENG)	87
Current affairs (weekly)	216
Entertainment	
Panel game	96
Sitcom	352.5
Sport	90

£1.00 = $1.50

Close to 80% of viewers watch television at least once per evening, with an average peak rating of 18%. The audience share is as follows:

ITV	45%
BBC1	37%
BBC2	14%
Channel 4	4%

Most viewers are satisfied with the programme mix. Most believe that television makes their lives more enjoyable. Those who are not entirely satisfied would prefer more plays, documentaries, and comedy, and a majority of viewers would still like to see more channels on offer. The top-rated programmes are quiz shows and continuous soap operas, while news/current affairs, light entertainment, plays, and films are popular. Of the programming 14% is foreign, much of it from the United States.

BBC1 and 2 are funded through licence fees as they are non-commercial stations, while the ITV's major source of funds is advertising. The ITV companies pay 20% of their income to the IBA. This is distributed as follows: 6% to the IBA on rentals paid by the independent companies for use of the transmitters, 12% to fund Channel 4, and 2% as a corporate levy to the government.[19] Advertising is broadcast daily on ITV, Channel 4, and TV-AM during natural breaks in the programmes throughout the day. An average of 6 minutes per hour is allowed. Commercial airtime can be purchased on a national or regional basis. Requests for commercial airtime is submitted 3 to 4 months in advance. The CPT for ITV is $9.9 for a peak-time 30-second commercial slot. The IBA has its own code of standards on advertising which must be followed by the regional companies. The United Kingdom has the most commercial airtime available in Europe, and supply currently meets demand.

The BBC operates four domestic non-commercial radio services: Radio 1, 2, 3, and 4. Programme content is as follows: Radio 1—pop and rock music; Radio 2—light music, information, and sport; Radio 3—serious music, culture and educational programmes; Radio 4—spoken word, current affairs, and music. The weekly audiences of Radio 1, 2, 3, and 4 are 47, 40, 9, and 22% respectively of the national audience. Figures exceed 100% as more than one station may be listened to by a single individual.

Commercial radio is operated under private broadcasters who have contracts with the IBA. There are currently 38 local stations which are allowed to broadcast up to 9 minutes per hour of advertising. Radio advertising revenues exceed $123 million annually.

2.5.16 WEST GERMANY

Broadcasting in West Germany is regulated by the eleven regional Länder governments. Within each Land, a broadcasting organization has been established and a Radio and Television Board and Programme Advisory Board are responsible for compliance by broadcasters with the guidelines set forth.

The Zweites Deutsches Fernsehen (ZDF) provides national programming, while the Arbeitsgemeinschaft der Öffentlich-Rechtlichen Rundfunkfernanstalten Deutschlands (ARD) consists of a network of nine regional stations that transmit local programming in the early evening and national transmission during the remainder of the time. In addition, various ARD stations combine to operate five regional outlets, forming a third television channel that is non-commercial. The regional stations are as follows:

Bayerischer Rundfunk	BR	Munich
Hessischer Rundfunk	HR	Frankfurt
Norddeutscher Rund-funk	NDR	Hamburg
Radio Bremen	RB	Bremen
Saarländischer Rundfunk-Fernsehen	SR	Saarbrucken
Sender Freies Berlin	SFB	Berlin
Süddeutscher Rundfunk-Fernsehen	SDR	Stuttgart
Südwestfunk	SWF	Baden-Baden
Westdeutscher Rund-funk	WDR	Cologne

There are also five separate regional networks for the areas served by BR, HR, SDR/SWF, and NDR/RB/SFB.

The ARD was established in 1950 to represent the joint interests of the broadcasting corporations. The ZDF or second German television channel was established with an agreement between all Länder in June 1961. All of the stations are public corporations entirely independent of the government. Each corporation is governed by a Rundfunkrat, the Broadcasting Board, a Verwaltungsrat, the Administrative Board and an Intendant, the Director General. The Director General is responsible for daily operations of the channels, while the Broadcasting Board and Administrative Board are responsible for advising and management, including programme guidelines.

ARD transmits about 60 hours per week, while ZDF transmits over 85 hours per week. Regional programmes vary in transmission times. ARD has an average peak rating of 19%, while for ZDF the figure is 17%. The total television reach is 26.5%. News is the largest programme category on ARD, while culture and education are the largest category of programming on ZDF. Of the programming 15 to 20% is foreign, much of it American. A complete breakdown of programming is listed in Table 2.7.

About 25% of the viewers can receive foreign programmes from France, Belgium, Netherlands, Denmark, Austria, and Switzerland. Only 30% of the viewers are satisfied with the programming choice and 39% are satisfied with the quality. The average viewer watches 1 hour 21 minutes per day of ZDF and 1 hour 19 minutes per day of ARD. Germans are growing increasingly dissatisfied with television and are watching less than they were one year ago. Many viewers complain that there are too many political programmes, and the 'upper classes' tend to look down upon television.

Funding for the ARD and ZDF national channels is provided through

Table 2.7 German programming

ARD*	Programmes by type (%)
Drama	18.7
Light entertainment	10.3
Music	2.8
News	30.2
Sport	7.3
Films	12.9
Daily reviews and women's	11.6
Other	6.2
Total	100

ZDF†	Programmes by type (%)
Culture and education	16.8
TV films and plays	19.0
Documentaries	2.4
Light entertainment	7.6
Theatre and music	5.3
News	11.3
Politics and current affairs	10.5
Sport	6.2
Morning programmes	11.5
Continuity	5.6
Advertising	2.4
Other	1.4
Total	100

* From ARD.[21]
† From ZDF.[22]

a combination of licence fees and advertising (40% for ZDF, 36% for ARD). The licence fee is 16.25 DM per month and contributes $1796 million towards operating costs annually. Advertising revenues contribute over $517 million to operating costs. The third regional channels are non-commercial.

Advertising is broadcast daily on both ARD and ZDF except for Sunday. There is a limit of 20 minutes daily per channel, and it is transmitted in blocks. The blocks are usually scheduled close to popular programming, which is often American. The blocks are less popular with up-market viewers than with the general population. ARD's blocks are during the local programming period of 18.00 to 20.00, while ZDF's blocks are between

17.30 and 19.30. Of the nine ARD stations, seven sell advertising individually, while SDR and SWF sell together. The ARD has carried advertising since 1954. The ZDF carries national advertising and has done so since 1964.

Commercial airtime for the following year is purchased in September, and only the month can be specified. There is considerable oversubscription for available commercial time. The average CPT for ARD is $3.2 and the average CPT for ZDF is $2.5.

The regional organizations carry advertising provided by companies directly affiliated with them and designed specifically to sell broadcasting time. The ZDF also directly manages its sales. Advertising broadcasts are subject to controls by the broadcasting or television boards.

German radio consists of regional ARD stations operated by the Land. Each Land operates three channels. About two-thirds of the stations are commercial, and WDR is the only Land without at least one commercial station. The national reach of all of the combined commercial stations is 16.4%. Radio advertising revenues exceed $209 million annually.

The Land of Lower Saxony is considering local commercial stations. The first are planned for Hanover in 2 to 3 years and will be commercial.

ZDF is participating in 3-SAT, a German language general entertainment service distributed via satellite to cable homes. ARD has taken an option on an Intelsat transponder to deliver a variation on its main programme: ARD-Plus.

2.6 Conclusion

Broadcasting in Europe has evolved under strict governmental control, and private broadcasting services are limited. Most of the broadcasting organizations have been granted a monopoly by the state to operate. A government board usually has ultimate control over the schedule and content of the programming.

Most broadcasting organizations operate one to two channels on a national basis. Where regional or local channels exist, they usually collaborate to form a national network, as in the United Kingdom (ITV), France (FR3), and Germany (ARD).

Programme content varies from one country to the next, but much of it is of an informative nature. This is particularly true in the Scandinavian countries whose channels are nearly all devoted to education and information. Much of the general entertainment programming is imported from the United States and is often found among the most popular television programmes.

Programming is provided in four major language groups: English, German, French, and Scandinavian. Several countries produce programming in several languages, including Belgium, Finland, and Switzerland. Nearly every country is capable of receiving foreign programming and this programming reaches audiences of significant size, particularly in Switzerland.

There is considerable interchange among the Scandinavian countries with similar languages, and superb linguists, such as the Dutch and Belgians, receive and view the most foreign programming. Countries such as Greece and Portugal have little reception outside their own countries because their languages are not widely understood.

Most broadcasting corporations are financed by a combination of advertising and licence fees. While commercial television exists, it is very limited, and most countries are oversubscribed for advertising time. Several countries, such as Denmark, Belgium, and Sweden have no commercial television whatsoever. Advertising is often shown in blocks at specific intervals during the evening; it rarely interrupts a programme, except in the United Kingdom, and the minutes per day of commercial airtime is relatively limited. There are many varying restrictions on products that may or may not be advertised. There is very little flexibility with airtime purchase and, in many countries, commercial time is bought over one year in advance.

Nearly every broadcasting corporation is undergoing change. Traditionally restrictive legislation is becoming more flexible, allowing the broadcasting systems to expand or become involved with new media ventures. Commercial television is becoming more acceptable and is recognized as a necessary source of revenue with the increasing competition for viewers. It is apparent that old barriers are breaking and the broadcasting industry will become extremely diversified over the next few years. As in the United States, it is the publishing groups who are entering the television media through their participation in cable and television relay programmes.

References

1. Home Office, *Direct Broadcasting By Satellite*, Her Majesty's Stationery Office, London, 1981, p. 32.
2. Stephen Birkill, 'The EBU and Eurovision', *Satellite Orbit International*, August 1984, p. 34.
3. 'EBU Active Members', *EBU Review, Technical*, February 1985, p. 61.
4. Commission of the European Communities, *Realities and Tendencies in European Television*, Commission of the European Communities, Brussels, 1983.
5. John Parry, 'EBU wants Olympic rights reduction', *Advertising Age*, 30 July 1984, p. 39.
6. John Parry, 'EBU wants Olympic rights reduction', *Advertising Age*, 30 July 1984, p. 34.
7. European Communities, *Treaties Establishing the European Communities*, Articles 59 and 60, Office for Official Publications of the European Communities, Netherlands, 1978, pp. 268–269.
8. Alastair Tempest, 'The Witches' Brew', *Journal of Advertising*, Volume 1, 1982, p. 151.

9. John Howkins, Neville Hunnings, and Joanna Spicer, *Satellite Broadcasting in Western Europe*, International Institute of Communications, London, 1982.
10. J. Walter Thompson, *Television Today and Television Tomorrow*, J. Walter Thompson, London, 1984.
11. Euromonitor Publications Ltd, *Advertising in Western Europe*, Euromonitor Publications Ltd, London, 1984.
12. Saatchi and Saatchi Worldwide, 'Index of advertising expenditure per adult in Europe', *Marketing*, 5 July 1984, p. 84.
13. D'Arcy MacManus Masius, *Advertising Age's European Media and Marketing Guide*, Focus Magazine, London, 1985.
14. The details of broadcasting by country are derived from numerous sources. The primary sources include:
    Advertising Association, *Marketing Pocket Book 1984*, The Advertising Association, London, 1984.
    Campaign and Young and Rubicam, *The Advertiser's Guide to European Media*, Marketing Publications Ltd, London, 1983.
    Commission of the European Communities, *Interim Report: Realities and Tendencies in European Television: Perspectives and Options*, European Communities Publications, Brussels, 1983.
    D'Arcy MacManus Masius, *Advertising Age's European Media and Marketing Guide*, Focus Magazine, London, 1985.
    Euromonitor Publications Ltd, *Advertising in Western Europe*, Euromonitor Publications Ltd, London, 1984.
    J. M. Frost (Ed.), *World Radio TV Handbook*, Billboard Ltd, London, 1984.
    Information et Publicité en Europe, *Audio-Visual Media in Europe*, Information et Publicité en Europe, France, 1983.
    J. Walter Thompson, *Television Today and Television Tomorrow*, J. Walter Thompson, London, 1984.
    Ogilvy and Mather Europe, *The New Media Review*, Ogilvy and Mather Europe, London, 1984.
15. *Structure de Programmes des Chaînes Nationals*, UER et Organismes de Télédiffusion, France, 1980.
16. Oy Yleisradio Ab, *Facts and Figures 1984*, Oy Yleisradio Ab, Helsinki, Finland, 1984.
17. MTV Oy, *MTV Oy 1983*, MTV Oy, Helsinki, Finland, 1983.
18. SSR, *The SSR Handbook*, SSR, Zurich, 1983.
19. IBA, *Television and Radio 1984*, IBA, London, 1983.
20. Broadcasting Research Unit, *A Report from the Working Party on the New Technologies*, British Film Institute, London, 1983.
21. ARD, *ARD Handbook*, ARD, Frankfurt, 1983.
22. ZDF, *ZDF Handbook*, ZDF, Mainz, 1983.

3

Advertising in Europe

3.1 Introduction

Cable and satellite channels will expand the role for television advertising. Currently, European governments have been able to apply the rationale that the television airwaves are a limited resource and should therefore be used in the best national interests. Television content is regulated and the educational, cultural, and informational capabilities of the medium have been emphasized in addition to, or in some countries rather than, the popular entertainment dimension (discussed in Chapter 2). Funding is through a combination of a licence fee imposed on viewers and limited opportunities for advertising. Thus, broadcast television, by and large, is not viewed as a purely commercial consideration.

There is not, however, the same willingness by governments to pay for these new cable and satellite channels out of the public purse, either by way of indirect taxation or through direct revenue-collection methods, such as licence fees, as there was to establish public broadcasting. Most cable and satellite channels are being developed with a market philosophy—someone (not the government) has to pay. Either the user pays a subscription for the service, the programmes are supported by sponsorship (where permitted) or advertising revenue pays for the channel, or, in the case of some of the new public broadcasting consortia, the channels are cross-subsidized out of mainstream broadcast revenues.

Some governments, for example, the United Kingdom, Ireland, Luxembourg, and the Netherlands, are leaving the financing of 'new media' development to the private sector. In the United Kingdom, the government has stated that all of the costs for cable and DBS must be borne by entrepreneurs. In such cases, subscriptions, advertising, and sponsorship will be the sources of funding for cable and satellite services.

The dilemma facing governments in the next decade as the cable and/or satellite potential begins to be realized is whether to apply the same standards and constraints on advertising to these new media, and thus maintain a uniform code of practice, or to respond to the market pressure to relax advertising restrictions. The former will restrict growth and possibly allow for further concentration of power by existing broadcasters. The latter may encourage a diversity of entertainment-oriented cable channels which in turn could force traditional broadcasters to compete, i.e., to become more entertainment oriented and to seek a greater share of the advertising dollar by lobbying for greater freedom to collect advertising revenues. As pan-European services evolve, the potential exists to redistribute advertising revenue among European countries, from those viewing the programmes towards the countries originating transmissions.

The resolution of this dilemma is complex as broadcasting has been seen by most European governments as more than just entertainment—it is integral to the whole social and cultural fabric of the society. However, the solution is not solely within the hands of the national governments. Satellite communications from abroad, both direct to the household and to cable head ends, will provide viewing and advertising opportunities over which little or no effective national controls can be exerted.

Whatever moral position one wishes to take on advertising and media, the harsh realities are that most newspapers and magazines could not exist on subscriber revenue alone and that the limited choice available to television and radio audiences in Europe would be even more restricted without advertising. New media, such as cable and satellite, will not survive without advertising support. A key concern will be to identify where the potential is for new advertising dollars.

3.2 The European advertising picture

Europe may be considered as a total advertising market, a market with 355 million people and advertising expenditures of $24 billion (Table 3.1). The importance of advertising within European economies has been rising steadily over the last decade—but not to the same degree as in the United States. Whereas advertising in the United States has been growing rapidly,[2] the picture across Europe is one of constraint. In 1983, for example, the advertising share of the gross domestic product in the United States was 2.7% compared to a European figure of 0.8%. Put in per capita terms, this becomes $370 per head to $67. Thus, as a total advertising market, Europe must be viewed cautiously. Clearly, a potential exists.

Within Europe, the advertising picture is as different among various countries as it is between the United States and Europe. First, consider total advertising expenditure (Table 3.1). It ranges from 1.79% of total GDP in Finland to 0.23% in Portugal. It is correlated with population size, the countries falling into four natural groupings. Generally, countries with larger advertising expenditures allocate a greater share of their GDPs to advertising.

Table 3.1 Total advertising expenditure and GDP, 1983[1]

Country	Total advertising expenditure ($ million)	Advertising expenditure as a percentage of GDP	Total advertising expenditure rankings	Advertising expenditure as a percentage of GDP rankings
United Kingdom	5 838	1.27	1	3
West Germany	5 226	0.79	2	8
France	2 921	0.56	3	11 (equal)
Italy	2 043	0.58	4	10
Netherlands	1 807	1.37	5	2
Spain	1 389	0.87	6	7
Finland	886	1.79	7	1
Switzerland	850	0.80	8	6
Sweden	672	0.73	9	9
Norway	579	1.05	10	4
Denmark	573	1.01	11	5
Belgium	453	0.55	12	13
Austria	340	0.51	13	14
Greece	133	0.38	14	15
Ireland	101	0.56	15	11 (equal)
Portugal	47	0.23	16	16
Total	23 858			

The relative individual prosperity within nations, as measured by per capita GDP, is also correlated with per capita advertising expenditure (Table 3.2). It is interesting, however, to divide the countries into those with a high advertising expenditure per head (greater than $60) and those with less than $60 per capita:

1. For those countries with a per capita advertising expenditure of less than $60 there is a very high correlation with the per capita GDP (a correlation coefficient of 0.81). This suggests that for the relatively poorer countries of Europe, the per capita advertising expenditure is proportionally low.
2. For the richer countries of Europe, there is no observable association between the two measures, i.e., while the countries with per capita advertising expenditure in excess of $60 per annum are also those with the higher GDP per head, there is no statistical correlation between these two measures. The implication is that there is an observed 'flattening-off' or saturation for advertising across European countries linked to affluence. There is a point where advertising saturation of available media occurs and advertisers are not prepared to increase their budgets

Table 3.2 Per capita advertising expenditure, 1983

Country	Advertising expenditure per head ($)	GDP per head ($)*	Advertising expenditure per head rankings	GDP per head rankings
Finland	182	10 096	1	6
Norway	139	13 095	2	2
Switzerland	132	15 170	3	1
Netherlands	124	9 055	4	8
Denmark	110	10 858	5	4
United Kingdom	103	8 123	6	11
Germany	85	10 563	7	5
Sweden	80	10 916	8	3
France	53	9 441	9	7
Belgium	45	8 190	10	10
Austria	45	8 947	11	9
Spain	36	4 080	12	14
Italy	35	6 031	13	12
Ireland	28	5 025	14	13
Greece	13	3 416	15	15
Portugal	4	1 969	16	16

* From Bureau of Statistics of the IMF.[3]

further proportionally to economic growth. As three of the eight 'wealthiest' countries in Europe do not permit any television advertising whatsoever, it is reasonable to assume that per capita advertising expenditures would rise in these countries if television advertising were permitted. That is, there would not be just a substitution effect from, say, press to television, but the total expenditure on advertising would also rise.

Further insight comes from considering the relative national shares of total European advertising expenditures (Table 3.3). The United Kingdom consumes the greatest 'share' of advertising expenditures, accounting for 24.5% of all European advertising, while providing only 16% of the European population. Together with Germany, the two countries account for just under 50% of the total advertising expenditure for a combined population which is 34% of the total European population.

By cumulating the relative national shares of advertising and population, we see that:

1. The United Kingdom and West Germany account for 33.6% of the population and 46.4% of advertising.
2. The United Kingdom, West Germany, France, Italy, the Netherlands,

Table 3.3 Europe as a total market—national shares of advertising expenditures and population, 1983[1]

Country	National advertising expenditure as a percentage of total European advertising expenditure	National population as a percentage of total European population	Cumulative shares	
			Advertising (%)	Population (%)
United Kingdom	24.5	16.07	24.5	16.1
West Germany	21.9	17.57	46.4	33.6
France	12.2	15.47	58.6	49.1
Italy	8.6	16.18	67.2	65.3
Netherlands	7.6	4.08	74.8	69.4
Spain	5.8	10.81	80.6	80.2
Finland	3.7	1.37	84.3	81.6
Switzerland	3.6	1.79	87.9	83.3
Sweden	2.8	2.37	90.7	85.7
Denmark	2.4	1.46	93.1	87.2
Norway	2.4	1.17	95.5	88.3
Belgium	1.9	2.83	97.4	91.2
Austria	1.4	2.17	98.8	93.3
Greece	0.6	2.79	99.4	96.1
Ireland	0.4	0.99	99.8	97.1
Portugal	0.2	2.88	100.0	100.0
Total	100.0			

and Spain account for 80% of the population and 80% of the advertising expenditure.
3. France, Italy, the Netherlands, and Spain account for 46.6% of the population and 34.2% of the advertising.
4. Belgium, Austria, Greece, Ireland, and Portugal comprise 11.7% of the population and only 4.5% of the advertising expenditure.

Finally, consider a breakdown of total national advertising expenditure into television, press (magazines, trade press, and newspapers), direct mail, and other (radio, outdoors, cinema) (Table 3.4). The national allocations are influenced strongly by government policies restricting the extent of television advertising. Thus, press remains the dominant advertising medium in Europe—but as we show below, it is not expanding its share.

Table 3.4 Advertising expenditure shares across categories, 1983[1]

	Percentage allocations			
Country	Television (%)	Press (%)	Direct mail (%)	All other (%)
Greece	53.9	35.8	1.8	8.5
Portugal	43.5	24.0	1.5	31.0
Ireland	32.9	43.5	3.6	20.0
Italy	32.0	42.0	5.5	20.5
Austria	27.8	45.0	9.1	18.1
United Kingdom	25.1	50.6	9.9	14.4
Spain	23.3	38.4	7.2	31.1
France	15.5	48.2	8.6	25.7
Germany	9.9	70.5	11.6	8.0
Belgium*	9.5	58.6	18.0	13.9
Finland	8.5	68.0	10.8	12.7
Switzerland	7.2	64.2	5.0	23.6
Netherlands	4.5	57.9	30.0	7.6
Sweden	0	57.6	38.4	4.0
Denmark	0	56.0	33.0	11.0
Norway	0	70.2	24.0	5.8
Europe	16.3	55.6	12.7	15.4

*Includes revenue from RTL.

3.3 Television advertising

Four distinct groupings of countries emerge, according to the level of television advertising (Table 3.5):

1. Countries with no television advertising
   Scandinavia (Sweden, Norway, Denmark)
   Belgium (but significant revenues from RTL, spillover service from Luxembourg)
2. Countries with minimal television advertising (less than 20% of total advertising budget)
   Finland
   Central Europe (France, Germany, Switzerland, Holland)
3. Countries with 'average' share of total advertising budget from television (between 20 and 40% of total advertising)
   English language countries (United Kingdom, Ireland)
   Italy (with competitive private networks)
   Austria
   Spain

4. Countries with a significant share of advertising revenue from television (greater than 40% of the total advertising budget)

Less wealthy countries where television is *the* medium for communication (Portugal, Greece)

Revenue sources for the government-run television stations are derived from a combination of licence fees and advertising (discussed in Chapter 2). In nearly every country, a tariff is levied upon a television receiver set which is used to produce programming and fund the organization. In some countries, such as Denmark, Sweden, Norway, and Belgium, the licence fees are the sole source of funding (Table 3.5).

Advertising on television is very restricted, and it is often confined to blocks. Restrictions on specific products vary from country to country, but most disallow advertising of cigarettes, tobacco products, and alcoholic drinks and restrict advertising directed towards children. In their annual report on European new media, 1984, Ogilvy and Mather go so far as to suggest that advertising is merely 'to help fund public broadcasting'.[4]

Table 3.5 Television revenue in Europe†—advertising share of total revenue (advertising plus licence fees), 1983. (Countries are ranked by percentages of total revenue derived from licence fees.)*

Country	Estimated revenue from radio and television licence fees ($ million)‡	Advertising revenue from radio and television ($ million)	Advertising share of all broadcast revenue (%)
Italy	0	817.1	100
Spain	0	466.9	100
Greece	0	79.1	100
United Kingdom	1134	1465.3	65
Ireland	43	33.2	44
Portugal	32	20.4	39
Finland	139	75.4	35
France	889	452.8	34
Austria	245	94.5	28
Switzerland	154	61.2	28
Netherlands	244	81.3	25
West Germany	1796	517.4	22
Belgium	238	43.0	15
Sweden	256	0	0
Denmark	195	0	0
Norway	119	0	0

*Derived from Table 2.2 and Euromonitor Publications Ltd.[1]
† Excludes government subsidies to television networks.
‡ Some of the revenue from licence fees is used to support the radio networks.

Clearly, the effect of government restrictions has ensured that television as a source of revenue for broadcasters and an opportunity for advertisers has not been fully exploited in Europe. This may be about to change. Legislation has been passed in Belgium permitting advertising and it is likely to take effect in 1985. The Dutch New Media Law proposes to allow foreign channels to carry advertising, as long as it is not directed at the Dutch population. The various Länder in Germany are introducing more liberal laws for cable programmes, including those originating in Germany, and the Bredin Report in France also indicates that the new network would generate revenue through advertising. The British government has mooted that the BBC may not continually pass on all increased costs to consumers by raising the licence fee and a committee headed by Professor Alan Peacock has been set up to examine alternative sources of funding for the BBC, which include advertising and sponsorship, and their effects on broadcasting as a whole. The advertising scene in Europe is changing—new media and cross-national programming are at the core of the change.

3.4 Competing media: press and television

In the major European countries, press still remains the dominant advertising medium. There has been a general shift away from press towards either television or direct mail for advertising over the last five years (Table 3.6). It is now universally recognized that television is the best mass market medium for advertisers but, for targeted markets, it competes with direct mail and specialist magazines.

Consider the trends in the last five years between press and television. There were only three countries in which the relative share of press advertising increased—Finland, Switzerland, and Spain—but in only two of these countries was this at the expense of television. In Greece, Ireland, and particularly Italy there was a marked decrease in the share of advertising through the press which was almost completely counterbalanced by an increase in the television share. Relative share changes across the rest of Europe were not quite so pronounced:

Press	*versus*	*Television*	
Big loser ($<10\%$)	*and*	Big winner ($>9\%$)	Greece, Ireland, Italy
Loser (0–10%)	*and*	Winner (0–9%)	Austria, Belgium, the United Kingdom, France, the Netherlands, Portugal, West Germany

| Loser (0–15%) | *but* | No television advertising | Denmark, Sweden, Norway |
| Winner (0–5%) | *and* | Television unchanged | Spain |

Table 3.6 Changing share of national advertising expenditures[1]

| | Press share | | | Television share | | |
| | | | Change over | | | Change over |
Country	1978 (%)	1983 (%)	5 years (%)	1978 (%)	1983 (%)	5 years (%)
Austria	56.8	49.5	−7.3	25.0	30.9	+5.5
Belgium*	64.3	58.7	−5.6	6.2	9.5	+3.3
Denmark‡	60.6	56.0	−4.6	Nil	Nil	—
Finland	62.6	66.7	+4.1	9.5	8.8	−0.7
France	61.0	56.0	−5.0	14.5	18.0	+3.5
Greece	48.0	37.5	−10.5	47.0	56.5	+.9.5
Ireland	60.0	45.0	−15.0	24.0	34.0	+10.0
Italy	59.0	42.0	−17.0	20.0	32.0	+12.0
Netherlands	60.8	58.0	−1.2	23.9	30.0	+6.1
Norway‡	82.3	70.2	−12.1	Nil	Nil	—
Portugal†	27.2	24.0	−3.2	41.7	43.5	+1.8
Spain	49.7	50.9	+1.2	30.3	30.4	+0.1
Sweden‡	61.1	57.6	−3.5	Nil	Nil	—
Switzerland	61.4	64.2	+2.8	8.9	7.1	−1.8
United Kingdom	66.0	62.2	−3.8	26.6	30.9	+4.3
West Germany	72.6	70.5	−2.1	9.4	9.9	+0.5

*Television advertising from RTL, Luxembourg.
†Figures for 1981 and 1983 only.
‡For Denmark, Norway, and Sweden, the positive side of the equation has been. (See Euromonitor Publications Ltd.[1])

3.5 Constraints on television advertising

The introduction of cable and DBS will provide additional advertising opportunities, especially in the television markets where advertising is severely limited. The extent of new media advertising will depend upon the latitude built into the new media policies being formulated in most countries. Regardless of the vested interests of broadcasters and governments' collective desires to 'not-quite-deregulate' national airwaves, the harsh economic reality of the eighties is that someone will have to pay for the programming.

To put this newly emerging advertising market into perspective it is necessary first to describe the current state of television advertising in terms of its limitations for advertisers. Then, and only then, can we address the market opportunity which is created by this hiatus of regulations.

Three general variables have been identified which together describe the legal constraints on television advertising:

1. Maximum volume of advertising permitted—measured in minutes per day
2. Periods during which advertising is forbidden
3. Flexibility for the advertiser

The point of viewing the market from within this constraint set is that it establishes the scene for identifying future opportunities which may arise as a result of relaxation of constraints. By far the most significant constraint is the limited time available for advertising. It is changes in this restriction that will bring about most opportunities for cable and DBS networks.

3.5.1 TIME AVAILABLE FOR ADVERTISING ON BROADCASTING TELEVISION
The total television advertising minutes per day ranges from 640 minutes in Italy to 20 minutes in Austria, excluding Sweden, Denmark, and Norway which permit no television advertising (Table 3.7).

Table 3.7 Television stations carrying advertising

Country	Number of stations carrying advertisements	Total advertising time per day (min)†	Estimated average advertising expenditure per available minute ($)	Television advertising expenditure per capita ($)
Italy	7	640	2 798	11.2
France	3	54	22 970	8.3
Switzerland	3	60	2 794	9.6
Finland	2	25	8 258	15.5
West Germany	2	40	35 440	8.4
Greece	2	110	1 780	7.1
Ireland	2	83	1 099	9.3
Netherlands	2	30	7 427	5.6
Portugal	2	135	412	1.9
Spain	2	100	8 867	8.3
United Kingdom	2	140	28 675	26.1
Austria	1	20	12 952	12.6
Belgium*	1	68	1 735	4.3

* The revenue is from RTL.
† From Ogilvy and Mather Europe[4] and Association of Advertising Agencies.[5]

Three important observations can be made. First, with Italy as the exception, the total television advertising minutes permitted per day is low. Austria, France, Finland, West Germany, the Netherlands, and Switzerland all have less than one hour of advertising per day. In the United States, for example, the three national networks provide 540 minutes of advertising daily—up to 20 times as much as in some European countries.

Second, the United Kingdom is the only country in which the per capita television advertising expenditure is in excess of $20. Again, in the United States, the figure in 1983 was $57. Noting that television advertising expenditure per head is linked to national prosperity, Switzerland, France, West Germany, and the Netherlands have relatively low levels of per capita expenditure. A third and interrelated observation is that the advertising expenditure per available minute indicates not only the relative differences in the size of the viewing audience but also the restrictions on the supply of available television spots. This is most noticeable in Germany where both the low advertising expenditure per capita *and* the high expenditure per available minute are direct consequences of the restricted availability of airtime. There is a similar situation in France and the Netherlands.

The relative situation across European countries can be shown clearly by using the United Kingdom as a benchmark, i.e., supposing the United Kingdom is the closest to what an economist would term 'a free market' for advertising, where the available minutes and the per capita expenditure are such that there are no major shortages of prime-time advertising minutes or an excess of advertisers demanding these minutes.

Then, with respect to national television advertising expenditure as a share of GDP, all European countries are lower than the United Kingdom, even though Finland and the Netherlands spend a greater proportion of GDP on all advertising than does the United Kingdom. If we group countries we have:

50% or more of UK level	Austria, Finland, Greece, Ireland, Italy, Spain
Less than 35% of UK level	Belgium, France, Germany, Netherlands, Portugal, Switzerland

Given that the United Kingdom in turn is lower than the United States, the significant observation is that under a certain set of market and economic conditions, the total level of television advertising expenditure across Europe could rise substantially. As with other indicators, the countries which are the largest and more affluent have the lower relative expenditures and, of course, offer the greatest potential for advertisers.

3.5.2 TEMPORAL RESTRICTIONS ON ADVERTISING PLACEMENT

There are three general restrictions on times during which television advertising is permitted in Europe (Table 3.8). These are:

1. Limitations on advertising during certain times of the day, e.g., after 22.00 at night
2. Days of the week (or religious holidays) on which there is no advertising, e.g., Sundays
3. Limitations on the positioning of adverts, e.g., no breaks during programmes

Table 3.8 Time restrictions on advertising[6-8]

| Country | Limitations | | Advertising during programmes? |
	Times	Days	
Austria	Not after 22.00	Sundays, legal holidays	No
Belgium*	None	None	Yes
Finland	None	24, 25 December	Constrained; at natural breaks
France	None	May 1, after 20.00	No†
Greece	None	Holy week	Constrained
Ireland	None	None	Yes
Italy	None	Good Friday and 2 November (RAI only)	No (RAI only)
Netherlands	Blocks before and after news programmes	Sundays, Good Friday, Ascension, Christmas	Yes
Portugal	None	None	Yes
Spain	None	None	Yes
Switzerland	Not after 21.30	Sundays, 5 legal holidays	No
United Kingdom	None	None	Yes
West Germany	Not after 20.00	Sundays, 16 legal holidays	No

* Advertising forbidden on national channels. Above applies to RTL.
† Does not apply to the new Bevlusconi-Seydoux channel where advertising during breaks is permitted.

The most significant limitation is the restriction which prohibits programme breaks for advertisements in France and the German-speaking territories—Austria, Switzerland, and West Germany.

Table 3.9 Flexibility for the advertiser[7, 9]

	Lead time for placing adverts	Programme sponsorship
Austria	1 October for the following year. Adverts transmitted in blocks of 4 to 5 minutes between programmes	Limited; no adverts during programme
Belgium	Three months in advance. Adverts transmitted in 20 blocks per day. Average 6 minutes per hour	No
Finland	Two to three months in advance for each of the three selling campaigns per year (spring, summer, autumn). Screened in 'natural' programme breaks	No
France	Advertisers submit annual requests in October and schedules in November	No
Greece	Requests submitted first half of each month for preceding month	No
Ireland	Selling begins on a first-come basis 1 to 3 months before transmission	No
Italy	On private television stations adverts sold according to availability. Annual campaigns on RAI	No on RAI; some on private
Netherlands	Adverts transmitted in blocks. Applications by 15 September for following year. Only month of transmission can be specified	No
Portugal	Two weeks is the practical minimum	No
Spain	Five days' notice. Adverts in blocks, maximum 12 adverts per block and 4 minutes of adverts per hour	No
Switzerland	Adverts transmitted in 5 blocks of 4 minutes. Orders for adverts submitted by October for following year	No
United Kingdom	Requests made 3 to 4 months in advance. Periods of light demand—advert time bought during week it is to be transmitted	No
West Germany	Adverts transmitted in 3 to 5 blocks during period 18.00 to 20.00. Advertisers apply for time each September and can only nominate a month for their adverts	No

3.5.3 FLEXIBILITY FOR ADVERTISERS

There are three measures of the flexibility (or lack of flexibility) for European advertisers on broadcast television:

1. The lead time for spot placement
2. The existence of programme sponsorship
3. The restrictions placed on various products

The findings are summarized in Table 3.9. In essence, programme sponsorship is not permitted across the national networks. The lead times required of advertisers in specifying their campaign and the constraints imposed upon them is directly related to the availability of advertising airtime. In Austria, France, the Netherlands, Switzerland, and West Germany, advertisers apply for space up to one year in advance. Further, in the Netherlands and West Germany, advertisers are only able to specify the month in which they would like their advertisements placed. As a consequence, new products are difficult to advertise and small businesses are often excluded as pre-payment is frequently required. Targeting audiences is also difficult as there is little flexibility as to when a specific advertisement will be aired. Despite these constraints, the demand for spots still far exceeds the supply of commercial airtime.

In Finland, Belgium, Ireland, and the United Kingdom, around three months' notice is required. In the United Kingdom and Ireland, supply and demand dictate the final time at which advertising space can be requested. In Greece, Portugal, Spain, and Italy, the lead time is less than one month and adverts are sold according to availability.

As to product restrictions (Table 3.10), these differ widely, although there is a general consensus which bans tobacco and cigarettes, prescriptions and other pharmaceuticals, religious and political groups, and severely restricting advertisements for beer, wine, and spirits, personal hygiene, and family planning products as well as adverts directed at children.

With the increased demand for advertiser-supported broadcast services and the increase in transborder flow of programming from the new media technologies, there has been a movement within Europe to standardize advertising restrictions. The EEC has established some basic advertising guidelines in their Green Paper *Television Without Frontiers*.[10] This topic is discussed in more detail in Chapter 10.

This paper suggests that there should be 'co-ordination of certain specific aspects of the law of the Member States regulating advertising on radio and television' which would allow programmes that follow EEC guidelines to be freely broadcast and rebroadcast throughout the community. The Commission has also recommended that each member state should authorize advertising by some broadcasters within the private sector. This would allow the state-run channels to continue to be funded through licence fees without redirecting already thinning funds. Further, the EEC believes that increased advertising would stimulate interest in goods and services and boost the economy in general, as it has done in the United States.

Two key concerns are not considered: first, that growing markets are not usually the result of advertising, although differing brand shares within markets are very sensitive to the level of advertising, and, second, that a free market for transborder advertising would stimulate economies from

Table 3.10 Product restrictions for advertising on European television[7, 9]

Products	Austria	Belgium	Finland	France	Greece	Ireland	Italy	Netherlands	Portugal	Spain	Switzerland	United Kingdom	West Germany
Tobacco/cigarettes	×		×	×	×	×	×	×	×	×	×	×	×
Cigars	×		×	×		×	×	×	×	×	×		×
Beer/wine	×	×	×	×	O	O	O	O		×	×		×
Spirits	×	×	×	×		×	O	O	O		×	×	
Medical services											×		
Prescriptions	×	×			×		×	O		×	×	×	×
Pharmaceuticals	O	×	O				×	O		×	×		×
Family planning	×				×								×
Personal hygiene				O	O		O						
Gambling				O	×					×		×	
Money lending				O	×								
Political	×							×			×	×	×
Religious groups	×							×			×	×	×
Soliciting		×											
Jewellery				O									
Charities				O								×	
Dieting	O			×					O				
Toys				O									
Matrimonial							×					×	
Influence children	O							O		O			O
Entertainment				O									
Publishing				O									

× = product/product group legally forbidden
O = products subject to restriction

which the transmissions originated to the possible detriment of those nations receiving the programmes.

3.6 Advertising potential for new media

The restrictions and constraints on television advertising across Europe means that the television set is still relatively unexploited as an advertising medium. New media laws being drafted in most countries (see Chapter 10) allow satellite-delivered and terrestrial cable channels at least as much advertising time as the broadcast channels. In some cases, the amount of time is being increased substantially: e.g., in West Germany, the government is permitting cable programmes to carry advertisements for up to 20% of total airtime compared to 20 minutes per day on the broadcast networks. Further, as satellite-delivered channels cross national borders, local governments are conceding that their own broadcasting restraints cannot be seriously applied in their entirety. Liberalization of European airwaves is underway.

The US cable experience, however, offers a cautionary lesson. Cable advertising, even with 30% of households connected to systems, was less than 2% of total television advertising in 1983. The criteria for take-off as an advertising medium is the potential to deliver network-like viewing audiences.

In Europe, the satellite-delivered pan-European channels such as Sky Channel, Music Box, and TV-5 (see Chapter 8) have achieved substantial household penetration relatively quickly. In March 1985, for example, over 3 million households could receive Sky Channel.[11] This gives it a bigger market potential than the whole of Switzerland, Austria, Norway, Finland, Denmark, Greece, Portugal, and Ireland, and ranks it fifth in size among the ITV companies in the United Kingdom. The disadvantage, for advertisers, is that this large penetration is spread among many countries, eight in the case of Sky Channel. Still, the potential to give advertisers access to large audiences is now a reality.

There are different ways to estimate the television advertising opportunity across Europe. We consider a relatively conservative estimate of the potential as comprising the *immediate* unfulfilled demand for television advertising, i.e., the shortfall in supply at the current national prices.

3.6.1   UNSATISFIED DEMAND FOR TELEVISION ADVERTISING
Our intention is to estimate the additional television advertising expenditure which would accrue if outlets were available in national markets and overbooking was minimized. A number of assumptions are necessary:

1. In Greece, Ireland, Italy, Portugal, and Spain there is a relatively high share of total advertising on television now. Advertisers can organize campaigns at short notice and it is assumed that there is no significant shortage of opportunities for television advertisers. This does not mean that a new channel would not attract advertising revenue; nor does it mean that television advertising will not increase its share of all advertising. Our aim is merely to estimate the volume of advertising in Europe that does not find its way on to a television set even though the advertiser would like to advertise.

2. The United Kingdom is used as a benchmark for other European countries. There is no shortage of opportunities to advertise, and with Channel 4 and TV-AM, in addition to the ITV network, there is a two-tiered pricing system with 85% of the revenue in 1983 going to the

ITV network. The proposed benchmark is the ITV network component of total advertising, that is, 20% of total advertising.

3. There is no television advertising at present in Belgium, Denmark, Sweden, and Norway. In a 'free-market' situation, it is assumed that, because of the extended position of other advertising opportunities, television advertising would conservatively achieve 50% of the UK benchmark, that is, 10% of total advertising.

4. In the Netherlands and Switzerland, it is assumed that, in a more liberalized market, these nations would allocate a relative share to television advertising approaching 75% of the UK benchmark case, that is, 15% of total advertising.

5. In West Germany, the evidence from advertising agencies is that the unsatisfied demand for advertising about equals the volume of advertising on television at present. To be conservative, assume that in an unconstrained world, 29% of total advertising would go to television.

6. In France, the total advertising expenditure as a share of GDP is low. It is assumed that, with additional opportunities to advertise on television, both the total advertising share of GDP and the television components of total advertising would increase. It is assumed that, in France, 25% of the existing advertising expenditure could go to the television sector.

7. In Finland, although the share of total advertising expenditure on television is low (see Table 3.4), the per capita expenditure is relatively high

Table 3.11 Potential demand for television advertising

Country	Estimate of unsatisfied demand for television advertising ($ million)
Austria	0
Belgium	96
Denmark	57
Finland	0
France	278
Greece	0
Ireland	0
Italy	0
Netherlands	190
Norway	60
Portugal	0
Spain	0
Sweden	67
Switzerland	66
United Kingdom	0
West Germany	356
Total	1170

(see Table 3.7), as is the share of GDP allocated to advertising. It is assumed, therefore, that, in Finland, additional allocations to television due to unfulfilled demand are insignificant.
8. In Austria, there is a very restricted market opportunity for advertising generally, but television has a 28% share of total advertising expenditure. It is assumed that there is no significant unfulfilled demand.

Applying the above assumptions to European countries gives an estimate of unsatisfied demand for television advertising of $1170 million (Table 3.11) or an increase in television advertising in Europe of about 30%.

3.6.2 SUMMARY
In summary:

1. There is no broadcast television advertising in Sweden, Denmark, Norway, or Belgium. (Advertising in Belgium comes via spillover services, predominantly RTL, although legislation formulated in 1984 permitting advertising is likely to take effect in 1986.)
2. Italy is the only country in which there is any serious competition by broadcasters for advertising revenues. In all other countries a government or private broadcast monopoly is the only venue for advertisers. Flexibility is low, lead times for specifying campaigns is long, and in general demand exceeds supply for most of the year.
3. Of all countries allowing television advertising, Finland and Switzerland are the only two not registering an increase in television's share of total advertising allocations over the last 5 years. Italy, Ireland, and Greece recorded five yearly increases in television's share of total advertising in excess of 9%.
4. Austria, Finland, France, the Netherlands, Switzerland, and West Germany all allow less than 60 minutes of advertising per day.
5. Belgium, France, the Netherlands, Portugal, Switzerland, and West Germany all allocate less than 0.1% of GDP to television advertising.
6. Austria, Ireland, Italy, and the United Kingdom are the only countries in which per capita television advertising expenditure exceeded $10 in 1983.
7. Austria, France, the Netherlands, Switzerland, and West Germany require advertising campaigns to be prepared in the year prior to that in which the campaign will be aired. West Germany and the Netherlands only allow the advertisers (or the agency) to specify a preferred month for the transmission.

Central Europe and Scandinavia are relatively highly constrained, unexploited television advertising markets which are open to new consumer electronic media. For new satellite and cable channels, the key uncertainty is whether the public broadcasters respond to the perceived competition by appealing to governments for greater amounts of advertising airtime and thereby absorb some of this potential demand. (In 1984 the ITV companies launched a campaign to increase advertising airtime from 6 to 7 minutes per transmission hour but dropped the idea in 1985 when there was a down turn in revenue, and in Germany the networks proposed increases of 25%, from 20 to 25 minutes per day.)

3.7 Conclusion

Our aim has been to review the current position on broadcast television advertising in Europe, to highlight the differences and similarities across countries, and to determine if there is sufficient advertising potential to support new television channels. We are not advocating either more or less television advertising—merely contrasting and identifying potential opportunities for growth for new mass market media, such as cable and satellite.

Governments seem to have accepted that short of subsidizing the new channels out of the public purse, they will have to allow a more relaxed attitude to advertising. This is evident already in France—even under a socialist regime. In particular, there is the need to establish, either explicitly or by experience, a common policy for European-wide advertising.

The total advertising market is expected to grow, in real terms, and with it will come a growth in television advertising expenditures and opportunities. The difficulty for the new channels will be during the early years when the market penetration is low. In these times, it will be the absolute sizes of the audiences across national markets which will attract pan-European and national advertisers. The programming which has wide appeal or can be linked to identifiable target groups will win. The constraints on the existing broadcast media are such that a whole new strategy of advertising in response to market demands is possible.

References

1. Euromonitor Publications Ltd, *Advertising in Western Europe*, Euromonitor Publications Ltd, London, 1984.
2. 'Bates predicts slight rise in CPM', *Broadcasting*, 26 March 1984, p. 56.
3. Bureau of Statistics of the International Monetary Fund, *International Financial Statistics*, Volume XXXVIII, International Monetary Fund, Washington, DC, 1985.
4. Ogilvy and Mather Europe, *The New Media Review 1984*, Ogilvy and Mather, London, 1984.
5. Association of Advertising Agencies, *New Communication Developments in Europe*, Association of Advertising Agencies, London, 1983.
6. J. Walter Thompson Europe, *Television Today and Television Tomorrow*, J. Walter Thompson Co. Ltd, London, 1983.

7. Campaign and Young and Rubicam, *The Advertisers Guide to European Media*, Marketing Publications Ltd, London, 1983.
8. Economist Intelligence Unit, *Cable Television in Western Europe*, Economist Intelligence Unit, London, 1983.
9. Information et Publicité en Europe, *Audio-Visual Media in Europe*, Information et Publicité en Europe, France, 1983.
10. Commission of the European Community, *Television Without Frontiers*, European Economic Community, Brussels, 1984.
11. Kevin Cote, 'Commercial TV triumphs', *Focus*, April 1985, p. 9.

4

Cinema—An industry in decline?

4.1 Introduction

Cinema is one area of consumer media, which, on the whole, is in decline or at least in a state of non-growth throughout Europe. Reduction in the number of cinema seats, industry rationalization towards smaller cinemas, and declining attendances describe the European scene. The film industry is now highly concentrated—a small number of cinemas generate most of the box office revenue.

In no country in Europe is the per capita city cinema admissions equal to the level in the United States, where an individual visits the cinema on average 4.6 times per year and teenagers average one visit per month. On the contrary, annual per capita attendance has declined steadily since the boom times in the fifties when on average each individual in the United Kingdom visited the cinema 30 times per year. In 1982, the figure had fallen to less than 1.3 visits per person per year.

Across Europe, the cinema-going public is now predominantly the young, namely the 15 to 24 year old group. While they account for only 15 to 18% of the population, in most countries they make up over half of the cinema admissions.

Production of films in Europe is heavily subsidised by national governments and often closely linked with the production side of the national television networks. Consumer demand for local product is low, and without government support—in the form of financial assistance and the application of quotas on foreign products—most production would cease. The world demand for central European film production is also low, there being little interest by the American cinema, home video, cable, or television markets. Cultural and language differences provide additional barriers to the acceptance of this product.

4.2 Demand for US products

There is an increasing demand for US film products, even in relatively nationalistic countries such as France. This is not without some opposition as there is still some concern at widespread importation of foreign values and customs into most European countries.

Although comparative statistics are not kept on the relative viewing share of local and foreign films, it is still possible to get an indication of the importance of US film products. In Belgium, 31% of films released in 1981 were of US origin, about the same as those of French origin (33%).[1] The 1981 receipts from US films in Italy amounted to 33% of all box office receipts. Receipts from Italian films had declined from a high of 63% in 1972 to 44% by 1981.[2] The situation in Germany was even more significant. Using the distributors' share of box office receipts as an indicator, the share for US films increased from 31% in 1956 to 55% in 1984, while for German-made productions, the comparable figures are 47 and 11% respectively.[3]

In the United Kingdom, where Thorn and Rank together command 60% of the film audience, some 60% of all film products shown is US material. This share has tended to fluctuate since the mid-fifties, but during that time there has been a gradual decline in the share of UK films registered for cinema exhibition—from 23% in 1957 to 14% in 1981. Contribution to box office revenues of US film products is close to 85%.[4]

In France, the local product has declined as a share of all admissions, but was still 53% in 1982 (down from 58% in 1973) compared to the US share of 30% (up from 20% in 1973).[5]

In Austria and Belgium, the only recent data available on US film products indicate that, in 1982, around 40 and 30% respectively of all theatrical releases are of US origin.[6] These figures underestimate the respective share of box office receipts from US films, which are nearly double that number. In Switzerland, 48% of total admissions were for US films.

The net effect is that audiences across Europe are demanding more US film products. The special-interest, targeted national films, while guaranteed small audiences, are generally not major box office attractions.

The arrival of cable and DBS in Europe, even with minimal local content quotas, will require substantial amounts of new product. The challenge will be for the local industries to produce or co-produce the material demanded by these new home markets, or to 'import' a significant share of their culture from the United States.

4.3 Production and structure of local industries

A key problem for national film producers in Europe has been to find sufficient other national markets for the distribution of the local product. Language and cultural barriers make the lucrative US and Canadian markets relatively remote possibilities for significant revenue flows. Yet, as we have illustrated, the converse is not true. The French film industry, while consistently producing 150 to 250 films per year for the last decade, has little success in selling in international markets. The position is worse in Germany where even the local distributors and cinema owners are reluctant to book German films, preferring American films instead. The German film industry of late has been accused of catering primarily to minority audiences and as such has had little impact nationally and internationally. (Blockbusters of the fifties and earlier are still at the top of the most watched films in Germany.)

4.3.1 ESTABLISHED FILM INDUSTRY
There are three trends in local productions in all countries. First, an increasing number of films are co-productions, the co-producer being from another country. This trend has been stimulated by the increasing global market opportunities for the product, the 'loopholes' in local content requirements, and the tax advantages for both groups. Second, the national television networks are playing an increasing role in the production of films. Third, government support at some level seems to be necessary to sustain a viable industry.

Numerous government incentive schemes exist to stimulate local production. In France, for example, there is automatic assistance for producers of feature films as well as selective advances.

The advances accounted for 22.8% of the total French investment in feature films in 1982. The selective advances are repayable if the film is sufficiently successful at the box office. France has two other 'national' sources of revenue for film production: a cinema tax of around 13.5% imposed on every cinema admission and a special contribution from the three television companies in return for the 'privilege' of showing films on television. (This is in addition to the rights payments negotiated with producers and distributors.) In addition, the French government announced in 1985 that it was offering tax shelters for investors placing risk capital in new film or television productions—investments which would be tax deductible.

In Germany, the sources of contributors for an average film are:[3]

Private producer funds	10%
State film promotion fund	30%
BMI grant	12.5%
FFA loan	15.0%
FFA reference claim	17.5%
German television pre-sale	12.5%

The BMI (Bundesministerium des Innern) within the Ministry of the Interior makes grants, awards, and subsidies for all ranges of films. Criteria are artistic excellence, not commercial viability or even viewing appeal to national citizens. The FFA or German Film Fund Institute (Filmförder-ungsanstalt) is a public corporation under the authority of the federal government to promote the German film industry. Funds come from a levy on box office receipts and are used as subsidies, film aids for production, and promotion (repayable interest-free loans).

As in France, there is a commitment by the two German networks (the Film/Television Agreement) to assist in co-productions and subsidize projects. Films made under the terms of the agreement are given a cinema release of 6 months before appearing on television.[7]

The impact of television is probably the single greatest contributor to the decline of the film industry in smaller countries and to the decreasing popularity of the cinema. Not only is television a viable alternative as a source of entertainment, but television networks have begun to offer a wide variety of feature films. In some cases there is no restriction on the amount of foreign material that can be screened and no significant time periods between cinema and television release of the film product, i.e., 'audience gets what it wants when it is available!'.

In Austria, over 140 films were premiered on Austrian television in 1980 and a further 260 to 300 were shown on television. The number premiered equals half the total number of theatrical releases, and there are no restrictions on how soon after a theatrical release a film can be shown on television. There is some government support for the Austrian film industry, which produces about 8 to 10 films per year. The Federal Ministry for Education and Arts makes some money available to subsidize innovative directors at the OFF (Österreichischer Film Förderungsfonds—Austrian Film Promotion Fund), which is the main source of support.[8] It allocates around $20 million each year to provide:

1. Non-repayable project development grants (of up to $20 000) to films of high cultural importance
2. Interest-free loans to assist in the production of films and shorts which will raise the quality of Austrian films
3. Non-repayable grants to individual film makers
4. Support to help with the distribution of films.

In 1981, an agreement with the Austrian Broadcasting Authority (ORF) was worked out to encourage co-productions between the ORF and the OFF. Under the agreement, the ORF contributes around $20 million annually.

In Belgium, by contrast, television had little or no impact on the film industry in the fifties and sixties as it was virtually non-existent. Even today, the television networks buy foreign films rather than invest locally.

In the sixties, the Minister of Culture in Belgium introduced selective subsidies for film production and this provided the boost necessary to establish a commercial industry capable of producing 10 to 20 full-length films per annum. The subsidy does not encourage commercialism within film making and the producers tend to make low-budget, up-market, and basically uncommercial films. The positive side to this is that film making is not expensive in Belgium as there is no expectation of super profits. An average film costs $1.3 to $1.7 million.

In French-speaking Belgium, the Minister of Economic Affairs (Ministère des Affaires Economiques) gives post-production subsidies to Belgian full-length films shown in Belgian cinemas. (To qualify as a Belgian film, the film maker must be Belgian, shooting must take place in Belgium unless location shooting is required, and 50% of salaries must be paid to Belgian citizens.) A maximum of $450 000 is awarded per film. The Ministry for the French-speaking community (Ministère de la Communauté Française) gives subsidies and grants to promote Belgian films—production grants, script development, supplemental subsidies. The annual budget is around $2.2 million for these grants.

The Ministry of Flemish Culture has a similar budget, around $1.86 million, for investment in film production. It allocates much of this to short features, and to selected feature films it awards up to one-third of its budget.[1]

A totally different position exists in Switzerland. Here the film industry is financed by non-Swiss funds. Government support accounts for around 5 million Swiss francs—most of which goes into the production side (70%) and the rest into distribution. As the average cost of a full-length film is in excess of 1 million francs, and there are on average 10 to 12 films made each year, over 50% of financing comes from outside Switzerland.

Quotas on the import of foreign films into Switzerland, even for cinema release, are still applied, but even so less than 2.8% of total admissions in 1982 were for Swiss films.

Italian film production has declined from a peak of 250 films per year to a figure close to 100 to 120. The emergence of private television stations and their reliance on American film and series product is one of the main reasons put forward. Co-productions account for another 20 to 30 films each year.

The principal source of formal support is the SACC (Sezione Autonoma per il Credito Cinematografica) of the Banca Nazionale del Lavoro. It is a non-profit-making body (set up in 1935 under the Fascist rule) to promote the Italian film industry through the granting of medium-term loans with preferential interest rates. SACC also lends money against the promise of government assistance. The majority of Italian films are financed by SACC, 95 out of 114 in 1982, although the SACC loan is only available once production has started and is at most 60% of the film's budget.[2] Legislation in 1984 increased the amount of state aid to the cinema, a tax shelter for reinvestment of profits from production was set up, and VAT on film tickets was lowered.

The United Kingdom is also different. American companies provide between 70 and 80% of the finance invested in British production and, with the exception of Rank and EMI, all the major distributors are American. Government involvement in the industry has been limited. Until 1984, when it was phased out, there was a levy (the Eady Levy) on box office profits to direct some of the profits to producers, and the National Film Finance Corporation was set up in 1947 to make loans for films made for theatrical release. This latter organization, although effective in the early fifties has only around $2.2 million available per annum.[9] Alternatives have been discussed by the British government to replace the Eady Levy and a Films Bill has been proposed to provide support for the British film industry by imposing a levy on pre-recorded video cassettes and blank videotapes as well as on feature films screened by the BBC and Independent Television companies.

The industry is dominated by two companies—Rank and Thorn EMI—the only organizations concerned with distribution, exhibition, and production. The successful films are produced for the international market (i.e., America) or with other markets in mind—television, cable, and video. Recent successful films have all had some American investment in them.

The early eighties has seen the introduction of a number of independent production companies with large resources and access to financial investors from the City of London, e.g., Goldcrest and United Media. The film packaging aspects of the industry—structuring, financing, organizing—are key activities of the independents.

Goldcrest, the former Pearson-Longman subsidiary, has been restructured and raised $26.4 million in share capital early in 1984. With its links through Premiere into UK cable, recent successes on the overseas market (*Chariots of Fire*, *Local Hero*, *Gandhi*) and co-production opportunities, Goldcrest has made a major attempt to reposition the British film industry by announcing plans to produce feature films at a cost of $72 million in 1985.

The television networks have made a considerable amount of material in-house, but, until recently, they had little involvement in co-productions. The introduction of Channel 4 and the commercial realization of sales from overseas markets has encouraged some degree of co-production activity by the networks. In addition, independent producers such as Goldcrest are producing programmes which can be released both through cinema and television, which doubles their opportunity to earn revenues. They have also received a greater allocation of the budget from Channel 4 than they originally expected because they have produced programmes more cheaply than the ITV companies. The introduction of cable and DBS as additional

UK markets, plus creative financing from the City, may improve the situation even more in the eighties.

Thus, film production in Europe is in a state of flux: commercial pressures on the one hand versus cultural-artistic modes on the other. Without a large 'mass market', revenue is not forthcoming and investment dollars are hard to attract. Commercialization—and in particular for the America market—is not in the *modus operandi* of the artistic director producing for minority target markets. The United Kingdom, with a relatively weak pound and some support from the merchant banking fraternity, is beginning to attract American producers to UK studios. This does not change the cinema attendance trend in the short term, but may help to reestablish the production industry.

Commercial success in the eighties seems to be associated with co-productions, American investment, and multiple markets for product distribution.

4.4 Demand for cinema products

In the fifties, the cinema was unquestionably popular and was itself a large stand-alone edifice. Now cinemas are smaller and integrated within shopping areas and other accessible areas of high pedestrian traffic. This change is still in its earliest stages in many European countries, but the data indicate the trends clearly: smaller cinemas and fewer of them. For example, 70% of receipts came from 178 cinemas in Switzerland (37% cinemas), 87% of receipts came from 2300 cinemas in Italy (30% cinemas), and 60% of revenues came from just 39% of the cinemas in the United Kingdom, which are owned by just two organizations—Thorn EMI and Rank (Tables 4.1 and 4.2).

The cinema industry as a whole is characterized by falling attendances and rising prices. The cause and effect relationship is now lost in a downward spiral for attendance and an upward spiral for admission prices.

The changing patterns of cinema admissions is demonstrated clearly in

Table 4.1 Average cinema size (number of seats per cinema)[10]

| | Year | | | |
Country	1968	1973	1975	1981
Belgium	540		474	335
France		440	406	294
United Kingdom	997		574	397
West Germany	412		368	252

Table 4.2 Number of cinemas—selected countries[10]

Country	1957	1961	1968	1971	1975	1982
			Year			
Belgium	1585		773		562	475
Austria				770	626	536
France					4328	4709
Italy		10 441	9874	9324	8730	7726
Switzerland					562	506	477
United Kingdom	4194	2 711	1631	1482	1530	1432
West Germany			4060		3094	3598

Table 4.3 Cinema admissions in selected countries[10]

Country	*Admissions* (millions)			*Average annual change in admissions* (%)	
	1960	1970	1982	1960–70	1970–82
Austria	—	28.7‡	17.3	N/A	−4.5
Belgium	106.7*	32.42§	19.74	−7.6	−4.8
Finland	27.6	11.7	9.1	−8.2	−2.1
France	354.6	184.4	200.5	−6.3	+0.7
Italy	741.0	525.0	215.1¶	−3.8	−7.8
Netherlands	55.4	24.1	22.0	−8.0	−0.8
Norway	35.0	18.6	15.0	−6.1	−1.8
Portugal	26.6	28.0	30.3¶	+0.5	+0.7
Spain	—	331.0	107.0	N/A	−9.0
Sweden	39.6†	26.0	21.3¶	−5.8†	−1.8¶
United Kingdom	500.8	193.0	60.2	−9.1	−9.3
West Germany	605.0	160.0	124.5	−12.5	−2.1

Note: Where appropriate annual data are not available, alternative statistics used are: *1957; †1963; ‡1971; §1972; ¶1981.

Table 4.3. Of the 12 countries analysed, total annual admissions have fallen since 1960 in all but one country—Portugal, a country with limited and low entertainment-oriented television. The decline, however, in average annual percentage terms has not been constant over that period. It can be seen that the rate of decline in cinema admissions pre-1970 was much greater than in the post-1970 era.

The United Kingdom, Italy, and France are exceptions. The rate of decline in the UK has been relatively constant over the *whole* period. In Italy, it has even accelerated since 1970, the same time during which terres-

Table 4.4   Cinema admissions per capita[13]

Country	Year			Average annual change in per capita admissions (%)	
	1960	1970	1982	1960–70	1970–82
Austria	N/A	3.86	2.30	—	−4.2
Belgium	N/A	3.36	2.00	—	−4.2
Finland	6.23	2.54	1.89	−8.6	−2.4
France	7.76	3.63	3.73	−7.3	+0.2
Italy	14.25	9.78	3.77	−3.7	−7.7
Netherlands	4.82	1.84	1.56	−9.2	−1.4
Norway	9.78	4.79	3.65	−6.9	−2.2
Portugal	3.01	3.10	3.03	+0.3	0
Spain	N/A	9.80	2.85	—	−9.8
Sweden	5.29	3.23	2.56	−4.8	−2.0
United Kingdom	9.20	3.48	1.07	−9.3	−9.4
West Germany	10.92	2.63	2.02	−13.3	−2.2

Table 4.5   Decline in cinema attendance 1960–82[11]

Average annual decline in admissions	Period 1960–70	Period 1970–82
Rapid (9%+)	Germany, Netherlands, United Kingdom	United Kingdom, Spain*
Medium (4–9%)	France, Finland, Norway, Sweden	Italy Austria,* Belgium*
Low (4%−)	Italy	Germany, Finland, Netherlands, Norway, Sweden
Nil	Portugal	France, Portugal

* Data only available for 1970–82.

trial broadcasting has grown exponentially. In France, on the other hand, the decline in total attendances halted around 1970 and since that time there has been neither growth nor decline in cinema attendance.

Per capita attendance follows the above trends as population shifts have not been substantial in Europe (Table 4.4). The picture is simple: Europeans do not attend cinemas as frequently as they did in the sixties. The countries are grouped in Table 4.5 to show these trends.

4.5 Revenue from cinema products

There are two sources of revenue: admissions and advertising. By far the greatest source of revenue comes from admissions or box office receipts.

For comparative purposes, all European currencies have been expressed in US dollars. So that it is possible to observe changes in expenditure patterns, exchange rates for 1970 are used throughout. (To express each currency in its current dollar equivalent could compound real growth/decline for a particular country with local currency growth/decline relative to the US dollar.)

The exchange rates, taken from the *OECD Economic Outlook*, July 1984, are the average of daily rates. In 1970, $1.00 was equal to:

Austria	25.87	Schilling
Belgium	49.66	Franc
Finland	4.21	Mkk
France	5.53	Franc
Germany	3.65	DM
Italy	627.00	Lira
Netherlands	3.62	Guilder
Norway	7.15	Krone
Portugal	28.59	Esc
Spain	70.03	Ptas
Sweden	5.19	Krone
Switzerland	4.31	Franc
United Kingdom	0.42	Pound

4.5.1 BOX OFFICE RECEIPTS

Box office receipts have continued to grow over the last 20 years, although the number of cinema seats available and the size of cinemas have continued to decline. Two aspects of the growth in box office receipts are noticeable.

First, in inflation-adjusted units, the growth in box office receipts is not as dramatic except, again, in Portugal. Table 4.6 shows the growth without adjusting for inflation and Table 4.7 includes a CPI deflator. It is clear that the *relative* trends are not influenced by the deflationary index, i.e., inflation has had relatively similar effects across all countries. A cross-sectional analysis of average annual revenue growth shows that revenue growth is

Table 4.6 Box office receipts, 1960–82[14]

Country	Box office receipts ($ millions)			Average annual growth (%)	
	1960	1970	1982	1960–70	1970–82
Austria					N/A
Belgium		31.1‡	44.3		3.6
Finland	5.1	7.6	35.3	6.8	13.6
France	119.7	159.5	656.4	2.9	12.5
Italy	200.4*	290.1	716.1§	4.2	8.6
Netherlands	20.7	21.4	52.2	0.3	7.7
Norway	11.4	11.8	33.4	0.3	9.1
Portugal¶	6.84*	10.7	68.0§	5.1	18.3
Spain	N/A	94.1	389.2	N/A	12.6
Sweden	29.7†	34.3	86.2§	2.4	8.7
Switzerland	N/A	27.6	35.1§	N/A	2.2
United Kingdom	151.4	140.5	254.3	−0.8	5.1
West Germany	154.2	148.6	231.8	−0.4	3.8

Note: Where appropriate data are not available, alternative statistics used are: *1961; †1964; ‡1972; §1981.
¶ The Portuguese Esc fell dramatically against the US dollar from 28.6 Esc per dollar in 1970 to 79.4 Esc per dollar in 1982.

Table 4.7 Box office receipts—constant dollars‡, 1970–82[15]

Country	Box office receipts (millions of 1970 dollars)		
	1970	1982 (in 1970 units)	Average annual change (%)
Austria			
Belgium	31.1†	18.6	−5.0
Finland	7.60	9.9	2.6
France	159.5	206.0	2.6
Italy	290.1	161.7	−6.3
Netherlands	21.4	22.7	0.6
Norway	11.8	11.8	0
Portugal	10.7	24.2*	9.4
Spain	94.1	85.8	−1.0
Sweden	34.3	32.2	−0.8
Switzerland	27.6	20.4	−3.3
United Kingdom	140.5	58.1	−8.5
West Germany	148.6	126.8	−1.6

Note: Where appropriate data are not available, alternative statistics used are: *1972; †1981.
‡ The general national CPI deflator was used for all series and the 1970 average daily exchange rate with the US dollar is applied to express all series in the one currency.

not statistically correlated with increase in cost per admission or with the annual decrease in admissions. For example, in Finland, the decline in attendance has been more than matched by an increase in real (CPI-adjusted) revenues. Price increases in France have also outstripped declines in attendances, resulting in a real growth in box office receipts. Elsewhere, the rising prices in the seventies did not match the declining attendances and consequently there was an erosion of real total cinema earnings.

Table 4.8 Average cost per cinema admission, 1960–82. (Local currencies, unadjusted for inflation, expressed in 1970 dollars.)[16]

	Price of average admission ($)			Average annual change (%)	
Country	1960	1970	1982	1960–70	1970–82
Austria					
Belgium	N/A	0.72§	2.25	N/A	8.6
Finland	0.43	0.65	3.88	7.3	16.0
France	0.34	0.86	3.28	9.8	11.8
Italy	0.27*	0.55	3.32¶	8.2	17.7
Netherlands	0.38	0.89	2.37	9.0	8.5
Norway	0.32	0.63	2.22	6.9	11.0
Portugal	0.26‡	0.38	2.24	4.3	17.4
Spain	N/A	0.28	2.49	N/A	19.8
Sweden	0.67†	1.29	4.05¶	9.7	10.9
United Kingdom	0.31	0.74	4.21	9.1	15.6
West Germany	0.25	0.93	1.86	3.8	6.0

Note: Where appropriate data are not available, alternative statistics used are: * 1961; † 1963; ‡ 1964; § 1968; ¶ 1981.

Second, the average cost per admission has risen significantly while demand for cinema has fallen (Table 4.8). The post-war boom in cinema attendance and period of relative prosperity may have started the price spiral, but it would seem that now any causal link between rising prices and falling demand is difficult to substantiate. The correlation coefficient between the two series of annual average percentage changes in number of admissions and admission costs is not statistically significant, but it does indicate the negative trend.

In short, the net effect of the steep increases in admission costs has been that total box office receipts have continued to increase while attendances have declined.

4.5.2 CINEMA ADVERTISING

The revenue from cinema advertising is relatively small compared to box office receipts and is a minor component in total media expenditure, exceeding 2% only in Denmark (2.6%) and the Netherlands (5.4%) (Table 4.9). It has tended to fall off as the attendance figures have declined and

Table 4.9 Cinema advertising, 1981–82[17]

Country	Cinema advertising share of total media expenditure (%)	Cinema advertising ($ million)	Estimated revenue per thousand admissions ($)
Austria	0.4	1.2	69
Belgium	1.0	2.5	127
Denmark	2.6	1.6	N/A
Finland	0.3	1.8	198
France	2.0	44.3	222
Ireland	0.5	0.3	N/A
Italy	0.8	11.2	52
Netherlands	5.4	3.3	220
Spain	1.0	4.2	39
Sweden	2.0	4.2	197
Switzerland	1.3	7.4	N/A
United Kingdom	0.8	27.2	452*
West Germany	1.0	46.2	371

*In the United Kingdom there was an even greater than expected decline in admission in 1982. Even so, the advertising media expenditure is in general higher in the United Kingdom than elsewhere in Europe.

the demographics of cinema attendances have become youth oriented. Knowing the total admissions for a number of European countries, it is possible to represent the advertising expenditure in cinemas in a broad cost per thousand admissions measure. As an advertising measure, this figure needs to be linked to the number of advertisements, but it can be used to find an indication of relative expenditure by advertisers across media outlets.

4.6 Revenue versus expenditure

The matching of demand (admissions) and total revenue (box office receipts) relationships offers one way to identify general patterns in the cinema industry throughout Europe (Table 4.10).

There are countries where there is a stable, even growing cinema industry—Portugal and France—countries where the industry has failed to hold its own but not begun to decline dramatically—the Netherlands, Norway,

Table 4.10 The relationship of demand to total revenue 1970–82

	Growth in box office receipts		
Reduction in annual per capita admissions	*High:* 2 to 10%	*Low:* +2 to −2%	*Negative:* −2 to −10%
Low: just 2%	Portugal France	Netherlands	
Medium: 2% to −5%	Finland	Norway Sweden Germany	Belgium
High: −5% to −10%		Spain	United Kingdom Italy

Sweden, and Germany—and countries where all signs over the last decade are indicative of an industry in decline—the United Kingdom and Italy.

The product of the average number of visits to the cinema per annum and the average price per visit is an estimate of the annual per capita expenditure in constant units of currency—cinema as an item of personal consumption expenditure.

Table 4.11 Annual per capita expenditure on cinema (in 1970 US dollars)[18]

Country	1970 ($)	1982 ($)	Annual change over period (%)
Belgium	2.39*	1.89	−2.4
Finland	1.65	2.04	2.1
France	3.14	3.85	2.1
Germany	2.44	2.06	−1.7
Italy	5.40	2.83†	−7.0
Netherlands	1.63	1.61	0
Norway	3.03	2.87	−0.5
Portugal	1.18	2.42†	8.2
Spain	2.78	1.57	−5.6
Sweden	6.82	3.87	−6.1
United Kingdom	2.57	1.02	−8.8

* 1972
† 1981

It can be seen that in Table 4.11 that in only three countries—Portugal, France, and Finland—has there been a positive increase in cinema expenditure on the part of the average consumer. In Germany, Netherlands, Norway, and Belgium the individual has just failed to maintain constant annual spending patterns on cinema by reducing the number of visits to compensate for increased prices.

Cinema's relative importance as a discretionary expenditure item has declined due to competitive media, in particular television and VCRs, archaic cinema designs, and inaccessible locations.

In the United Kingdom, Italy, Sweden, and Spain, there has been a real reduction in the amount of per capita expenditure on cinema. The United Kingdom is the country where video cassette recorders have made the most

Table 4.12 Comparison of penetration and admissions[19]

Average annual admissions per capita	Penetration of colour television	
	Less than 50%	More than 50%
Less than 2.75		Belgium Finland Netherlands United Kingdom Norway
More than 2.75 visits	Italy Portugal Spain	Sweden France

impact and Sweden is not far behind. Italy has seen the mushrooming of all shapes and forms of private television stations. For the Spanish, who are ardent viewers of cinema and series entertainment on television and have only two television networks, the reduction is more to do with household economic factors than alternative competitive products.

We compared the penetration of colour television with the average annual cinema admission per capita (Table 4.12), believing that those countries in which colour television has made its greatest impact are those countries in which cinema attendance is lower. It is clear that where penetration of colour television is low, there is a relatively higher level of cinema attendance and where penetration of colour television is above 50%, cinema attendance is, on balance, lower.

4.7 Conclusion

No one factor is readily identifiable as the prime causal influence on the change in cinema attendances over the last 20 years. Each country has moved down a different path, although there are some commonalities:

1. Films in the eighties are aimed primarily at the young markets—15 to 24 year olds.
2. Cinema attendance is no longer a national pastime but rather a very select event; no country has an average annual per capita attendance of more than four visits.
3. The average size of cinemas is declining and in most countries there is also a reduction in the number of cinemas.
4. Europe has not followed the American trend, as yet, of integrating the cinema into shopping malls and other recreation/family centres although the trend is beginning, Milton Keynes in the UK is an example with a new complex of ten small theatres.
5. All television networks show a substantial number of films, even though in some countries release dates are up to 2 years after cinema releases.
6. Colour television and VCRs seem to have removed some of the incentive to visit cinemas frequently.
7. Annual per capita cinema expenditure has, in general, decreased steadily.
8. The demand for US film products continues to increase.

On the production side, the trend of joint production will continue and the role for television companies will become more significant especially with the growth in made-for-television movies and mini-series. For example, in July 1985, six European television organizations set up an association to produce high-quality programmes for Europe and elsewhere at a lower cost than any individual company could produce the product. The consortium consists of Channel 4 (UK), Antenne 2 (France), DRF (Austria), RAI (Italy), SRG/SSR (Switzerland), and ZDF (West Germany). It plans to produce 100 hours of drama over the period 1985–88 for a cost of 70 million ECUS (or around $55 million).[20]

Thus, television, with the enhancements of VCRs, cable and satellite, will become even more the all-consuming medium for entertainment unless the cinema repositions itself in the market.

References
1. British Film Institute, *Production Research in Western Europe—Belgium*, British Film Institute, London, 1984.
2. British Film Institute, *Production Research in Western Europe—Italy*, British Film Institute, London, 1984.
3. British Film Institute, *Production Research in Western Europe—Germany*, British Film Institute, London, 1984.
4. British Film Institute, *Production Research in Western Europe—UK*, British Film Institute, London, 1984.

5. British Film Institute, *Production Research in Western Europe—France*, British Film Institute, London, 1984.
6. *Variety*, 7 March 1984, 4th American Film Market Edition.
7. 'German speaking market at a glance', *Variety*, 7 March 1984.
8. British Film Institute, *Production Research in Western Europe—Austria*, British Film Institute, London, 1984.
9. Jeremy Tunstall, *The Media in Britain*, Constable and Company Ltd, London, 1983.
10. Statistics from the British Film Institute, London, Autumn, 1984.
11. Derived from the *Production Research in Western Europe* series.
12. Statistics from the British Film Institute, London, Autumn, 1984.
13. Calculations based on Statistics in [12] and Chapter 1.
14. British Film Institute, London, Autumn, 1984.
15. Based on [14] using national CPI deflators from the OECD Economic Outlook, No. 35, Paris, OECD, 1984.
16. Figures derived from previous calculations.
17. Campaign and Young and Rubicam, *The Advertisers Guide to European Media*, London, Marketing Publications Limited, 1983.
18. Calculations derived from earlier Tables and US currency exchange rates for 1970.
19. J. Walter Thompson Europe, *Television Today and Television Tomorrow*, London, J. Walter Thompson, 1983.
20. *Financial Times*, 'European Groups Get Together', 25 July 1985.

5

Direct broadcast satellites (DBS) and communications satellite systems

5.1 Introduction

The development of new consumer electronic media in Europe, be it cable or direct broadcast satellite (DBS), is contingent upon the availability of communications and direct broadcast satellites. Consequently, the position of satellite systems in Europe is crucial to the whole new media question.

The radio regulations distinguish between *fixed satellite services* (FSS), where communication takes place between fixed stations, registered with or approved by the PTTs; *broadcast satellite services*, where the transmission is intended for reception by the general public, reception of signals is unrestricted, and precise national guidelines (footprints, frequencies, orbital positions) are as specified by the World Administrative Radio Conference (WARC); and *mobile services*, where one of the end stations itself is mobile. The fixed and broadcast services are of major relevance to this discussion. First, however, we discuss the formal development of the European space industry.

Coordination of space activities in Europe has been under way for over 20 years. In 1960, ten nations formed the European Space Research Organization (ESRO) to develop experimental spacecraft for scientific purposes. Then in 1962, the European Launcher Development Organization (ELDO) was set up to provide satellite launcher capacity independent of the National Aeronautics and Space Administration (NASA) in the United States.[1] A third group, the European Space Conference, a committee of national ministers responsible for space affairs, was established in 1966 to set policies and guidelines for the European space effort.

This effort was fairly unsuccessful until 1973, when it was agreed to merge the three groups into one and set about implementing a coherent European space programme.

5.2 European Space Agency

In May 1975, the European Space Agency (ESA) took over from these three bodies with a mandate 'to provide for and promote, for exclusively peaceful purposes, co-operation among European States in space research and technology, with a view to their use for scientific purposes and for operational space application systems'.[2]

The ESA agreement was signed by 10 European countries (Belgium, Denmark, France, Germany, Italy, the Netherlands, Spain, Sweden, Switzerland, and the United Kingdom) with four others (Austria, Ireland, Norway, and Canada) having observer status. Ireland later signed and became the eleventh member.

In addition to its technological research programme and scientific programme, there are three major projects under way: Spacelab, a European satellite launcher, and a communications satellite programme. The situation in the European space industry is one of industrial and technological development first and then consideration of the market opportunity for any of the services which may be carried. There is no indication, at least in the consumer market, that any satellite programme or even initial planning decision has been based on consumer market needs.

The first project, Spacelab, the manned vehicle sent into space in 1983 aboard the NASA shuttle, has been successfully completed. Under the agreement, it is now the property of the United States for use by NASA. Further developments, including contributions to America's planned manned space station, are under consideration.

5.3 European satellite launcher — Ariane

The second project of the ESA is the European spacecraft launcher, Ariane. After an initial meeting in July 1973 of ministers responsible for space matters in the 10 European countries who were the original signatories to the ESA, there was an agreement signed in December 1973 between these nations and the ESRO to develop the Ariane satellite launcher.[3]

Equity was decided on national, political, and industrial interest in the programme as follows:

France	63.87%	Netherlands	2.00%
Germany	20.12%	Italy	1.74%
Belgium	5.00%	Switzerland	1.20%
United Kingdom	2.47%	Sweden	1.10%
Spain	2.00%	Denmark	0.50%

France, with 64%, was the major equity shareholder and, as such, the prime contractorship was awarded to its Centre National d'Etudes Spatiales (CNES).

87

The development phase for the Ariane launcher began in late 1973 with the ESA taking on the responsibility for monitoring and execution of the programme after 1975. ESA negotiated the launch contracts with users of Ariane and the industrial contracts for manufacture were placed by CNES. In 1978, the ESA member states decided to start production (manufacture and launch) of a 'promotion series' of six operational launchers (later reduced to four). This promotion series was designed to attract attention to Ariane and included construction of the payload facilities as well as a device for launching two satellites simultaneously.

After a successful first test flight in December 1979, Ariane's third test flight launched the first commercial satellites on ESA spacecraft, Meteosat-2 and an Indian communications satellite, Apple, in June 1981; the second test flight was a failure due to engine malfunction. A further test flight in December 1981 completed both the development and the promotion series, with three out of four successful flights.

In July 1980, the ESA decided to undertake further development of the launcher and commissioned a programme for two more powerful versions: Ariane 2 and Ariane 3. The objective of this programme was to increase the launch capability into the geosynchronous transfer orbit from 1750 to 2585 kg for a single launch or 2×1195 kg for a dual satellite launch.

By this time, too, it had been decided by ESA to create a private space transport company, Arianespace, to manage and supervise the production of Ariane, finance production, commercialize the launching service and control the launching operations function previously provided by the ESA. The shareholder participation in the company created in March 1980 includes 36 European manufacturers in the aerospace and electronic industries, 13 European banks, and CNES (Table 5.1). The national equity in Arianespace, which differs marginally from equity in Ariane, is broken down as shown below:[4]

France	59.25%	United Kingdom	2.40%
West Germany	19.60%	Sweden	2.40%
Belgium	4.40%	Netherlands	2.20%
Italy	3.60%	Denmark	0.70%
Spain	2.50%	Ireland	0.25%
Switzerland	2.70%		

Arianespace had its first commercial launch in May 1984, having now taken over that function from the ESA, and in mid-1985 was booked to launch 28 more satellites for 14 customers, including some in the United States.

Table 5.1   Corporate participation in Arianespace[4]

French shareholders 59.25%	*Swiss shareholders* 2.7%
Aérospatiale	CIR
Air Liquide	Controves
Comsip-Enterprise	F + W
CNES	Union des Banques Suisses
Crouzet	
Deutsch	*Spanish shareholders* 2.5%
Intertechnique	CASA
Matra	Sener
Saft	
Sep	
Sfena	*Swedish shareholders* 2.4%
Sfim	SAAB-Scanis
Sodeteg	Volvo
Crédit Lyonnais	
BNP	*British shareholders* 2.4%
Banque Vernes	Avica
Société Générale	Badg
Banque de Paris et des Pays Bas	Ferranti
	Midland Bank Ltd
German shareholders 19.6%	
Dornier	*Dutch shareholders* 2.2%
MBB/ERNO	Fokker
MAN	Allgemene Bank Nederland
Bayerische Vereinsbank AG	
Dresdnerbank	*Danish shareholders* 0.7%
Westdeutsche Landesbank Girozentrale	Rovsing
	Copenhagen Handelsbank
Belgian shareholders 4.4%	
ETCA	*Irish shareholders* 0.25%
Fabrique Nationale	Adtec
SABCA	Aer Lingus
Italian shareholders 3.6%	
Aeritalia	
BPD Difeza-Spazio	
Selenia	
Istituto Bancario San Paolo di Torino	
Bastogi Sistemi	

5.4 Communications satellites for cable programming

The third ESA project and the most significant for cable television is the communications satellite programme. (Direct broadcast satellites are discussed in Sec. 5.9.)

ESA began by organizing the construction of an orbital test satellite (OTS) programme. Its aim was to meet the telecommunication requirements for point-to-point communication, telephone, telegraph, and intra-European telex defined by the 26 European postal and telecommunications authorities which form CEPT (the Conférence Européene des Administrations des Postes et des Télécommunications) and to meet the needs of the European Broadcasting Union (EBU), a non-governmental organization to coordinate sound and television broadcasting services (Sec. 2.2). The OTS programme, in which one of two satellites was successfully launched in 1978, was the forerunner to the European communications satellite (ECS) series.[5] Two satellites in this series have since been launched successfully by Airianespace, Eutelsat 1–F1 (June 1983) positioned at longitude 13 degrees east and Eutelsat 1–F2 (August 1984) positioned at longitude 7 degrees east.

A further satellite, Eutelsat 1–F3 was destroyed at launch in September 1985 due to malfunction of the spacecraft. A fourth may be launched in 1986 as a replacement. The ECS spacecraft have all been designed and built under the management of the ESA by an industrial team comprising some 36 major companies in Europe headed by British Aerospace, Matra, and Selenia Spazio under the banner of Satcom-International.

The available frequency spectrum allows for a total of 12 channels for primary services and two channels for multi-services. The ECS satellites are identically configured with nine simultaneously operable 20 watt transponders each of 83.333 MHz bandwidth. Orthogonal linear polarization is used to maximize use of the bandwidth and minimize interference. The satellites have their uplinks in the 14.00 to 14.50 GHz frequency band and downlinks in the bands 10.95 to 11.20 GHz and 11.45 to 11.70 GHz for primary services (television relay and telephony) and 12.50 to 12.75 GHz (multi-services such as the national and international digital transmission of business services use TDMA as a prearranged basis at 64 K bits per second or preassigned basis at 2.048 M bits per second).[6]

The satellites have four coverage zones for the primary services—Spot West, Spot East, Spot Atlantic, and Eurobeam (Fig. 5.1). The multi-services are not available on Eutelsat 1–F1 but are offered on Eutelsat 1–F2 and will be offered on successive satellites. On Eutelsat 1–F1, five channels out of the nine are eclipse protected and all nine are to be eclipse protected on later satellites. The ECS satellite system is a multi-purpose one; its original stated aim was to provide telephony, telex, and data traffic among its members in Europe and to provide television relay services among broadcasters in the member countries of the European Broadcasting Union (EBU).

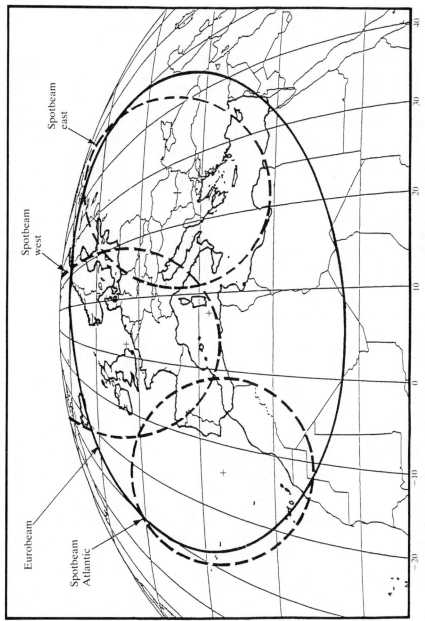

Figure 5.1 The four satellite coverage zones for the primary services (Copyright Interim Eutelsat).

Spotbeam east

Spotbeam west

Eurobeam

Spotbeam Atlantic

5.4.1 EUTELSAT

The satellites are owned and managed by Eutelsat, an organization formed to act for PTT administrations in establishing, operating, and maintaining the regional European telecommunications satellite systems.

The organization began as Interim Eutelsat in June 1977 with 16 of the 26 members of the CEPT. In September 1982 a definitive Eutelsat structure including the convention and operating agreements was outlined. By September 1985 this agreement had been ratified by 26 PTTs. Of the 26 signatories, 20 are shareholders (Table 5.2), the other 6 being Iceland, Liechtenstein, Malta, Monaco, San Marino and the Vatican City. The body was seen initially as a challenge to Intelsat's (International Telecommunications Satellite Organisation) global monopoly on the grounds of the 'economic harm' caused by a regional system to Intelsat, but the two organizations have since established cordial working relationships and Intelsat has made available its global infrastructure of satellite services to Eutelsat.

Eutelsat consists of:

1. An Assembly of Signatory Parties (the 26 nations) which lays down the principles and regulations for industrial satellite procurement
2. The ECS Council, the principal decision-making body responsible for developing a reliable space segment
3. The General Secretariat or executive branch

Eutelsat provides its member signatories with internal public telecommunications transmission of television programmes and business communications. The transmission of television includes a contract wih the EBU to transmit high-quality television programmes between EBU member broadcasting authorities through its lease of two transponders on Eutelsat 1–F2. It also includes the lease of transponders to signatories who in turn lease them to programme distributors.[4]

On Eutelsat 1–F1, for example, there are nine such leases through PTTs for cable programme delivery. The allocation of transponders was linked to the financial shares of member countries in the ECS space segment of Eutelsat (Table 5.2) as well as their stated demand for transponders.

Eutelsat has specified a pricing formula, depending upon whether the transponder is eclipse protected (or not) or whether the service is preemptible or non-preemptible. The first ECS satellite contains some non-eclipse-protected transponders which do not have batteries to operate them during the time of solar eclipses. Without solar power, they do not function. In addition, preemptible clauses relate to the rights of the lessee in the case of transponder failure. A totally preemptible transponder is one for which, if failure occurs elsewhere on the satellite, the lessee is preempted by another lessee. All transponders on Eutesat 1–F1 are preemptible.

Costs of a transponder excluding uplinking costs are (in ECU millions*):

Eutelsat charge to PTT signatory for transponders plus one national downlink	2.0
Eutelsat charge for additional downlinks; second to fifth at 0.2 million ECUs. No additional charge for further downlinks	0.8
	2.8

Table 5.2   Transponder allocation on Eutelsat series 1[7]

Country	Equity in Eutelsat (%)	Number of transponders		
		F1	F2	F3§
France	16.40	1		
United Kingdom	16.40	2		2
Italy‡	11.48	1		2
Germany	10.82	2		
Netherlands†	5.47	1		
Switzerland	4.36	1		
Belgium	4.92	1		
Austria	1.97			
Cyprus	0.97			1
Denmark	3.28			
Finland	2.73			
Greece	3.19			
Ireland	0.22			
Luxembourg	0.22			1
Norway	2.51		1	
Portugal	3.06			
Spain	4.64			1
Sweden	5.47			1
Turkey	0.93			1
Yugoslavia	0.96			
Total	100.00	9	1	9

* By 1982 there were 20 members.
† Transponder being used by EBU.
‡ Italy to vacate transponder once F3 is operational and Luxembourg to shift from east spot beam (spare) to the Italian west spot beam.
§F4 is to be launched in 1986 to replace F3 following the launch failure in September 1985.

The second ECS satellite is used primarily for telecommunications traffic, although the Norwegians are using one transponder for television relay within Norway. Transponder allocation may be modified in response to national demand for television relay transponders and up to five transponders may be made available for cable relay, on an interim basis. The intention is to shift these services to Eutelsat 1–F4 (the replacement for F3)

* At the time the contracts were being let in late 1983, 1 ECU was approximately equal to $1.00.

once it is launched and maintain two cable relay satellites, with a mixture of preemptible and non-preemptible transponders. The proposed transponder allocation is for F3, which it is assumed will carry over to F4, was:

United Kingdom,   2 (each)
  Italy
Norway, Denmark,   1 (each)
  Sweden, Spain
  (Atlantic Spot),
  Turkey (East Spot)

Although the programme has only just started, Eutelsat is already planning for the replacement satellites, Eutelsat 11, through the ESA. The request for proposals on the satellites was published in 1984 and then modified in 1985 as Eutelsat attempts to determine the optimal follow-up to its low-power system. These satellites are expected to be larger, up to 16 transponders, more powerful, 30 watts per transponder, and with European-shaped footprints. They are scheduled for launch beginning 1989.

As it has become clear that television services are the bread and butter of Eutelsat, a position that may not always be true as the data services markets evolve, in July 1985 Eutelsat working groups began to evaluate the feasibility of introducing a high-power satellite system (up to 100-watt transponders) as a complement to the present system. This project, although still in the planning stages, is being formulated as a broadcast system and not a fixed services system and is intended for launch some time from 1990 to 1992.

5.5 Other point-to-point communications satellites

Eutelsat is not the only source of satellite services for point-to-point broadcast relay services. The initial rush of applicants for the nine transponders on Eutelsat 1-F1, coupled with the planned development of cable in the United Kingdom and West Germany, encouraged the respective PTTs to lease unused transponders from Intelsat satellites. The beams from three transponders on Intelsat V-F10 at 27.5 degrees west have been refocused to fall on the United Kingdom and half-transponders have been leased by cable programme relay services Screensport, The Movie Channel (a pay-service which has gone into liquidation and been replaced by a movie/entertainment channel called Mirrorvision), and Premiere. Additional channels include Cable Network News (a full transponder) and Lifestyle (partial use of ScreenSports' transponder). Negotiations between Intelsat and Eutelsat have guaranteed that the Intelsat transponders can be used for international downlinking.

The Deutsche Bundespost has also leased six transponders from Intelsat.

The transponders are on Intelsat V located at 57 degrees east and are intended to stimulate the growth of cable in Germany. They have been allocated to various groups of Länder, in addition to ARD and offered to private cable programmers. By July 1985 there had been little positive response and only one lessee (German Music Box).

The French launched Telecom 1 in 1984, a multi-purpose telecommunications satellite with six channels in the KU-band (12 to 14 GHz) for business and video relay services, four channels at 4 to 6 GHz for telephone and television links within France and between overseas colonies and dependencies, and two channels at 7 to 8 GHz for fixed, mobile, and naval station defence services. Although the satellite has played no role in the development of cable in France or Europe, a second, launched in May 1985, is being marketed by the French for television relay services in Europe. The French negotiated a 'management contract' with Eutelsat for at least one transponder so that PTTs would approve its use for international downlinking. In October 1985, Satellite Television (STV), a private broadcaster, took the first lease on the system to provide a French entertainment channel.

Further satellites proposed in the fixed satellite services (FSS) spectrum (11 to 14 GHz) include Videosat, a French satellite for cable relay planned for the late eighties, Kopernicus, a similar satellite for Germany, and Italsat, a telecommunications satellite to be used for digital telephony, videoconferencing and experimental services. Two additional 'private' systems under development are the GDL project in Luxembourg and EBS in Sweden. The GDL project is an FSS medium-powered system which has the support of the Luxembourg government and institutional shareholders. Both projects are viewed as competitors to Eutelsat and DBS and are discussed in detail in Sec. 5.12.2. The success of the launch of these satellites will depend on the market situation and the regulatory breakthroughs which can be negotiated. Any private telecommunications satellite system in Europe proposing to offer international services must show that it will not cause economic harm to Eutelsat. Unless individual PTTs or groups of PTTs support the projects, or their use is redefined to be for national markets, Eutelsat acceptance will be slow. On the market side, slowdowns in national cabling programmes, lack of consumer demand, coupled with increased costs, may increase the riskiness of the ventures.

The ESA regularly produces tables of available transponder capacity on KU-band satellites in Europe and a summary position is produced below. It should be noted that not all orbital locations are favourable and the footprints, especially those for Intelsat, are not shaped for particular European counries. Drawing on the *ESA Bulletin* of May 1984,[8] and noting that some of the satellites have not been launched and may not get launched, the situation is shown in Table 5.3.

Table 5.3 Available transponder capacity on KU-band satellites for television services*

Launch year	Satellite	Number of available transponders assuming all proposed satellites are launched	
		Transponders	Cumulative number
Launched			
1982	Intelsat V-F4	3	3
1983	Eutelsat 1–F1	9	12
	Intelsat V-F7	3	15
1984	Intelsat V-F8	3	18
	Telecom 1A	5	23
	Eutelsat 1–F2	2	25
1985	Telecom 1B	5	30
To be launched			
1985	Eutelsat 1–F4	9	39
1986	TDF-1 (DBS)	4	43
	TV SAT-1 (DBS)	4	47
1987	GDL-1 (SES)	16	63
	Intelsat V1–F1	10	73
	Intelsat V1–F2	10	83
1988–90	Videosat–F1	12	95
	Kopernicus F1	10	105
	Irish (DBS)	5	110
	TDF–2 (DBS)	1	111
	Tele-X (DBS)	3	114
	Unisat (DBS)	3	117
Non-DBS transponders			97
DBS transponders			20

*This does not include any satellites launched by the United States for trans-Atlantic communication.

In addition to these European systems, four US-backed organizations independently filed with the Federal Communications Commission (FCC) in 1983–1984 for permission to launch and operate satellites to provide international business and video entertainment. These proposed satellites are direct competitors with Intelsat, although all argued that their services would complement rather than compete with Intelsat.

The first application was from Orion Satellite Corporation. It proposes to launch two KU-band satellites in late 1986 or 1987 and has filed with the International Frequency Registration Board (IFRB) to locate the satellites at 322.5 and 310 degrees. The aim is to develop a European market for American television programming and to offer trans-Atlantic video and data services for corporations. It is proposed that all capacity will be sold to private entities.

International Satellite Inc. (ISI) was the second group to apply to the FCC with a similar application. It, too, seeks approval to operate two KU-band satellites (302 and 304 degrees) but, unlike Orion, ISI plans to serve the video distribution and high-speed data markets on a tariffed, common-carrier basis for around 15 to 30% of the system's capacity. The principal stockholders in ISI are TRT Communications Inc. Co. (a subsidiary of United Brands)—43%, Satellite Syndicated Systems—15%, and Kansas City Southern Industries Inc.—14%. Private shareholders have the remaining equity in ISI.[9]

The third group, Cygnus Satellite Corporation, filed a proposal, similar to the one put by Orion, with the FCC in March 1984. They wish to launch two KU-band satellites (with a third on the ground) to provide video programming, private voice services for specific companies (but not public switched voice services, i.e., ordinary telephone services), and data services including videoconferencing, high-speed facsimile, computer communications, remote printing, teletext, videotex, and data-collection and distribution. They have also identified international DBS as a further service. It also proposes to file to provide a similar system in the Pacific region.

The fourth applicant is RCA Americom. They have approached the FCC to get permission to modify Satcom 6 to offer European beams. Satcom 6 is a C-band satellite authorized for launch in July 1987 to be located at 293 degrees. If their application is approved, 6 of the 24 transponders would be switched from domestic US coverage to provide beams over Western Europe. Again, the proposed markets are the teleconferencing and video relay.[10]

The key hurdles, in addition to getting financial backing for the projects and FCC approval, will come from Intelsat and, at least initially, the European PTTs. So far, the PTTs have been successful in controlling the lease, installation, and maintenance of telecommunications receiver equipment and have kept data communications and teleconferencing tightly under their control.

From the Intelsat perspective, the point of contention is Article XIV(d) which requires the coordination of national space segment facilities designed to meet international public telecommunications services requirements so as to avoid significant economic harm to the global system of Intelsat.[11] The response from the four companies is that they are not proposing to offer public switched voice services, only video and data traffic, and these latter services comprise only 15% of Intelsat's traffic.

The position in the United States has taken a step forward with the Presidential determination in November 1984 that 'separate international communications satellite systems are required in the national interest'. The

FCC began examining the issues raised by this determination in early 1985.[12]

Whatever the outcome, the French Telecom 1 is a first tentative move towards open-skies policies in Europe. It is a national system and will ultimately survive economically only if it offers services which could have been offered on Intelsat or Eutelsat satellites. In fact Eutelsat is managing one transponder for the French and that transponder is available for international use. Eutelsat has maintained control through cooperation. The British government's decision to give Mercury a similar status to British Telecom will promote competition and increase the likelihood of at least some support for one or more of the US projects in the United Kingdom. It is unlikely that carte-blanche approval will be available in the short term for all of Europe. Rather, if any US project becomes operational, downlinks will have to be negotiated on a country-by-country basis, just as they have been by the lessees on Eutelsat.

5.6 Satellite construction in Europe

The position on European corporate participation in the manufacture and construction is extremely complex. The companies who specialize in the manufacture of specific satellite components (solar arrays, antennas, amplifiers, multiplexes, switching, and telemetry and telecommand and control (TTC) subsystems) have tended to form different consortia and to be awarded subcontracts for the assembly of various Intelsat and Eutelsat satellites.

In 1966, a consortium calling itself MESH was established to respond to requirements of the ESA and was successful in its first bid for a space contract.[13] The group comprises:

British Aerospace	UK
Matra	France
ERNO	Germany
SAAB–Scania	Sweden
Aeritalia	Italy
Fokker	Netherlands
INTA	Spain

For the ECS series British Aerospace was the prime contractor with major industrial participation from:

Matra	France
ERNO	Germany
SAAB–Scania	Sweden
Aeritalia	Italy
ANT Nachrichtentechnik	Germany
Selenia	Italy

This group has been associated in full or part with the whole European family of communications satellites:

1. OTS
2. ECS
3. MARECS (the maritime telecommunications satellite launched in December 1981 and operated by INMARSAT)
4. SKYNET (for the British government)
5. EUROSTAR (the expanded ECS satellites being developed jointly by Matra and British Aerospace and proposed for the UK DBS project)
6. Telecom 1 (the French telecommunication satellite launched in August 1984 with Matra, the prime contractor, and Alcatel Thomson, the contractor for the telecommunications module)

A further permutation of this team has been commissioned by the ESA to construct a large multi-purpose telecommunications satellite (originally named L-SAT). This will carry an array of telecommunication and experimental services, including two channels suitable for home direct broadcast. The consortium is led by British Aerospace and includes Aeritalia (Italy), Fokker (Netherlands), Selenia-Spazio (Italy), and Spar Aerospace Ltd (Canada).

British Aerospace is also collaborating with Scott Science and Technology (USA) to design a satellite transfer vehicle for positioning large payload satellites in geostationary transfer orbit.

A second major satellite manufacturing endeavour emerged from the Franco-German effort CIFAS (Consortium Industriel Franco-Allemand pour le Satellite Symphonie). This consortium was established to launch the Symphonie satellites (August 1974 and December 1974) for cultural, educational, and experimental purposes. Unlike the ECS satellites, which operate in the KU-band (11 to 14 GHz), the Symphonie satellites use the 4 to 6 GHz bandwidth.

The French–German consortium had originally planned to be part of the ESA's L-SAT programme, but the costs, delays, and slow strategic development (pre-operational satellite and then to operational in the late eighties, early nineties) encouraged them to opt out (although French and German corporations are still subcontractors). They have chosen instead to set up a project management team comprising representatives of:

Centre National d'Etudes Spatiales (CNES)
Deutsche Bundespost (DBP)
Télédiffusion de France (TDF)
Deutsche Forschungs- und Versuchsanstalt für Luft- und Raumfahrt (DFVLR)

The work on satellite development is being carried out by an industrial consortium, Eurosatellite. Formed in 1978, Eurosatellite GmbH comprises Aérospatiale (France), Messerschmitt-Bolkow-Blohm (MBB) (West Germany), ANT Nachrichtentechnik (West Germany), which was the former AEG-Telefunken telecommunications division now 100% owned by Mannesmann, Bosch and Allianz Insurance, and ETCA-ACEC (Belgium). Each French and German company has 24% equity, ETCA being a minority shareholder with 4%.[14] Eurosatellite is developing, manufacturing, and delivering for launch a DBS satellite for France (TDF-1), Germany (TV-SAT), and Sweden (Tele-X).

5.7 Background to DBS

The International Telecommunication Union (ITU) has allocated fixed satellite services (FSS) and broadcast satellite services (BSS) to different frequency bands to avoid mutual interference and to maintain the distinction between broadcasting and intermediate transmission and telecommunication services.

The Intelsat, Eutelsat (ECS), and Telecom satellites are all in the FSS spectrum but the distinction is becoming blurred, as evidenced in the United States where technological developments have allowed lower powered transponders to achieve broadcast service standard performance with relatively small receivers. It is estimated that by the end of 1985, there are nearly two million private antennas receiving signals from the C-band satellites delivering programming to cable systems.

At the 1977 World Broadcasting Satellite Administrative Radio Conference, the rules and subsequent plan for national DBS services in Europe (region I) were outlined. Each nation was conceived of as wanting to offer direct reception services at some time and so an orderly allocation of five frequencies per country was adopted.

The services are to operate in the 12-GHz frequency band (11.7 to 12.5 GHz), the band that was allocated for broadcast satellite services for region I at the 1971 World Administrative Radio Conference (WARC).[15] The WARC '77 plan, which came into effect on 1 January 1979, divided the frequency spectrum into 40 channels and was designed to provide each European country with the potential of a DBS service using 0.9-metre receive antennas. The 40 channels are spaced at 19 or 18 MHz with a channel bandwidth of 27 MHz.

DBS was conceived of as a *national* service. Four orbital positions (5 degrees east, 19 degrees west, 31 degrees west, 37 degrees west) were allocated to the 25 countries of western and southern Europe (Table 5.4). The use of circular polarization (both left hand and right hand), 6-degree orbital spacing, and national channel allocations across the whole of the 800-MHz spectrum resulted in minimum technical interference and limited footprint overspill, while still ensuring each nation the frequency capacity to offer the equivalent of five television channels with multiple sound tracks. It is interesting to note that the original six members of the European Community, France, Germany, Italy, Belgium, the Netherlands, and Luxembourg, took the central European orbital allocation with good European-wide coverage, 19 degrees west, and the United Kingdom accepted 31 degrees west.[17]   .

The WARC '77 plan specified the signal power flux density (PFD) for

Table 5.4   WARC '77—orbital positions for European countries[16]

Orbital position	Frequency spectrum*	
	Lower half (11.7–12.1)	Upper half (12.1–12.5)
37° W	Andorra (L) San Marino (R) Lichtenstein (R)	Vatican (R) Monaco (R)
31° W	Portugal (L) Ireland (R) United Kingdom (R)	Iceland (L) Spain (L)
19° W	West Germany (L) Austria (L) France (R) Luxembourg (R)	Switzerland (L) Italy (L) Belgium (R) Holland (R)
5° E	Finland (L) Norway (L) Sweden (L) Denmark (L) Turkey (R) Greece (R)	Nordic (L) Sweden (L) Norway (L) Cyprus (R) Iceland (R)

*The letter in parenthesis indicates the assigned polarization. L = left-hand circular; R = right-hand circular.

the national footprints on the basis of assumptions about the receiving systems, including a G/T ratio of 6 dB based on an antenna diameter of 0.9 metres. This resulted in a maximum PFD of -103 dBW/m^2 at the edge of the footprint. Clear guidelines were set to reduce radiated power across neighbouring countries. The plan also made assumptions about the maximum antenna beamwidth, assumed to be 2 degrees, and required that the total interference at any receiver input satisfy an equivalent protection margin (a measure of the effective co-channel and adjacent channel interference).

The maximum radiofrequency power output ratings to achieve national footprints differed among countries and ranged from 40 watts for Luxembourg to 230 watts for Spain.

5.8 Technological developments and WARC '77

Technological developments since 1977, which have impacted the assumptions at the core of the WARC DBS plans for region I, have rendered the 1977 guidelines unnecessarily conservative. Improvements in antenna and receiver designs over the last seven years have resulted in increases in the efficiency of the receive antenna, reduced noise figures for LNCs, and the achievement of a 2-dB gain through threshold extension. In addition, better modulation techniques offer the potential of a further 2- to 3-dB gain.

For Europe, the implications of these developments which are close to 10 dB on the downlink budget are fourfold:

1. It becomes technologically possible to provide 'DBS-like' services from lower-powered satellites, including the ECS telecommunication satellites such as the Eutelsat series—the US experience revisited.
2. The comparative cost–revenue studies show considerable economic gain for a medium-powered configuration of an 800-watt payload (e.g., 16 × 50 watts) rather than a high-powered one (e.g., 4 × 200 watts).
3. Given the PFD specifications and the resultant national footprints, much smaller receive antennas are possible for the same picture quality.
4. As a corollary to implication 3, the feasibility of single national markets for DBS in Europe is questionable. Whereas the technological parameters in 1977 made access to national DBS overspill services unattractive because of the size of the antenna required to pick up the service from an adjacent country, this is no longer valid.

The most striking example of this potential was the results of a French test by the Centre Nationale d'Etudes des Télécommunications (CNET). Using Telecom 1, a satellite with 20-watt transponders, tests indicated that high-quality pictures could be received using 1-metre dishes. The engineers believe that it is feasible to use antennas with a diameter of 0.8 metres.[16]

Consider, for example, the position for WARC DBS in France. The 'overspill' footprint, i.e., the antenna beam with a PFD of -111 dBW/m², covers the United Kingdom, Ireland, most of Spain, Italy, the Netherlands, Belgium, Denmark, Germany, Yugoslavia, Austria, and Switzerland. The antenna diameter on the perimeter of this footprint to receive a signal with a carrier-to-noise of 14 dB is less than 0.90 metres—a challenge to the concept of *national* DBS and a boost to European DBS.

5.9 DBS systems in 1985

The reasons for the national pushes towards DBS are many and varied. France is seeking technological superiority in the aerospace industry, West Germany is aware of the industrial implications and is a willing co-partner in the Franco-German industrial development, the United Kingdom wanted to introduce additional competition as well as stimulate industry, and Ireland and Luxembourg could see economic implications if a major DBS or satellite project was coordinated from their respective countries.

Market demand and programme content considerations played a relatively minor role in the deliberations prior to 1985, the focus being primarily on the industrial and political ramifications for the national states. DBS also opened up the possibility of a new broadcast transmission standard after PAL to take advantage of the higher-quality pictures and sound. Competing standards were proposed by the French and British, both variants on a multiplexed analogue component system developed by the IBA[16] (see Sec. 5.12.5).

By 1985, 8 years after WARC '77, the DBS picture was anything but clear. The economics of the ventures have been challenged given the technological developments, their suitability for multiple programme packages is debatable, the national footprints are for most countries sufficiently restrictive as to preclude widespread access to international markets, and the technology of high-power travelling wave tubes has proved to be more difficult to develop than originally believed.

In this section, the status of each national project is reviewed briefly. As projects are still undergoing change, the aim is to provide a framework in which to observe trends and to follow developments over the next few years.

5.9.1 EUROPEAN SPACE AGENCY AND DBS

The European Space Agency (ESA) commissioned the 'L-SAT' satellite project to develop and demonstrate in-orbit large satellites with multipurpose payloads. The governments of Austria, Belgium, Canada, Denmark, Italy, the Netherlands, Spain, and the United Kingdom are participating in the project development while the industrial team for the satellite is led by British Aerospace (BAe).

The first satellite, to be named Olympus, is to be launched in 1988 and located at 19 degrees west. It will have two channels for DBS, in addition to low-powered transponders for other specialized services. The two high-powered transponders will be leased by Italy for pre-operational services and the European Broadcasting Union (EBU) for a pan-European television channel. The two projects will use one frequency allocated to the Italians and Austrians respectively. The EBU, which is deciding whether or not to run an advertising-sponsored entertainment/cultural/news channel, has been guaranteed 3 years of free usage of the transponder by the ESA, beginning in January, 1988. The Dutch, Irish, Italian, and German governments have agreed to experiment with the proposed channel by transmitting an advertiser-supported channel on the Dutch transponder on Eutelsat 1–F1 from October 1985.

The Italians have a similar arrangement. The national broadcaster RAI has taken a lease on a Eutelsat 1–F1 transponder as a means of developing a DBS service for Italy and to experiment with high-definition television. The arrangement with the ESA is structured so that the Italians do not pay for the transponder for the first 2 years and pay in the third year only if they intend to go ahead with their DBS plans. At present, the Italians still have plans to implement their own DBS system SARIT which is to eventually take over from Olympus. SARIT is based on a Eurostar platform with three television transponders. Selenia-Spazio will be the major manufacturer and a launch in 1988 is being planned.

5.9.2   UNITED KINGDOM

The original project envisaged that the British Broadcasting Corporation (BBC) would operate its own DBS satellite with 2×230-watt transponders and the ITV companies would operate a similar satellite—both satellites to be located at 31 degrees west (Fig. 5.2).

This proved not to be feasible as the BBC is limited in its abilities to raise funding on the open market, is precluded from advertising, and is forbidden to pass all of the costs onto the UK taxpayer by way of an increase in television licence fees. Subsequently, the Home Office proposed a three-channel tri-partite consortium for UK DBS and offered to withhold the allocation of the other two WARC channels for a period of 3 years, to allow the system to become established in a competition-free world. Thereafter, the Independent Broadcasting Authority (IBA) would be empowered to award 12-year DBS franchises in the same way it awards television franchises. The government endorsed the C-MAC transmission standard suggested in the Part Report.[18]

The tri-partite DBS venture comprised the BBC with 50% equity, the 15 ITV companies with 30% equity, and third-party companies, selected by the Home Secretary on advice from the IBA, with 20% equity. The venture was to operate under a board comprising four members of the BBC board of governors and four members of the IBA. The chief source of revenue was to be subscription payments, although the channels were permitted to carry advertising when and if the consortium members decided. It was a stated government objective that the project was not to be funded out of the public purse.

The third-party companies in the consortium chosen for their complementary skills to the BBC and ITV were Thorn EMI, Granada TV Rentals,

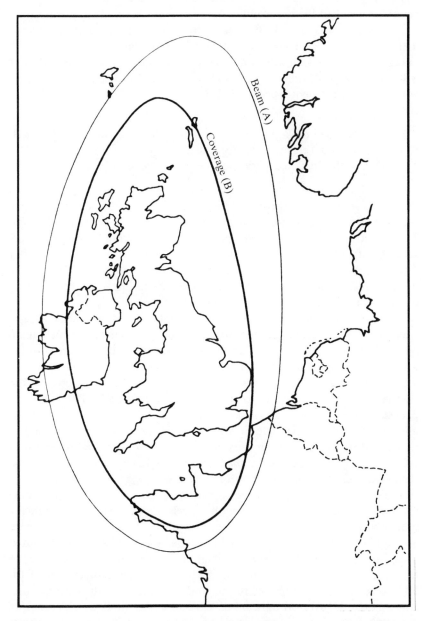

Figure 5.2 Unisat—UK broadcasting coverage. Contour A is a 3-dB beam edge contour for perfect pointing; B is the contour within which the clear weather power flux density is within 3 dB of the beam centre value. (From United Satellites Ltd.)

Virgin Group Ltd, Pearson plc, and Consolidated Satellite Broadcasting Ltd (whose major shareholders include Radio Télé-Luxembourg and a number of independent British producers).

The members assessed the project, its feasibility, and the possible services that could be offered. They accepted the basic structure of the high-powered satellite system but identified a number of modifications to original IBA or BBC proposals to minimize risk and increase the probability of an economically viable system. These included:

1. Reducing the annual amortized (or lease equivalent) cost of the satellite system from $95 million to $48 million, a reduction which United Satellites Ltd (Unisat) eventually agreed to, and
2. Extending the franchise period of the venture in excess of the proposed 10 years

The need for these concessions is demonstrated easily. For example, assume a $24 monthly subscription for three channels ($11 for programming and $13 for hardware) and a market penetration of 1.6 million households is achieved after 4 years. The negative cash flow would be $240 million after 3 years (compared to early estimates of $360 million or even $480 million by the consortium); the fourth year would be the first profitable year of operation. Payback would take another 3 years.[19]

The size of the negative cash outflow, the uncertainties of market acceptance of the service, and the technological changes that may render high-power DBS obsolete before it starts, caused considerable concern as to the wisdom of the venture. In April 1985, the BBC had its request for an increase in the television licence fee from £48 to £65 scaled down to £58 and felt under more pressure to cut back on developments. The ITV companies, concerned that the Peacock Committee of Inquiry, a committee set up to explore alternative funding arrangements for the BBC, might suggest that the BBC be involved in advertising or sponsorship, began to worry that their monopoly on television advertising might be reduced. This would then create financial uncertainty for their DBS hopes. There was also the realization that without a European market opportunity, the project would involve an element of competing with oneself—more costs with marginally more revenue.

The ITCA allowed the individual ITV companies complete discretion in deciding whether or not to participate in the DBS ventures. As an incentive to participate, the Home Secretary lifted the statutory requirement of the IBA to readvertise ITV franchises in 1989 and extended the existing franchises until 1997 (the withdrawal of a franchise is still at the discretion of the IBA). The statement was worded so that extension of a franchise was not contingent on participation in DBS, but it is understood that waiving the franchise was contingent on the joint DBS project going ahead.

United Satellites Ltd (Unisat), a joint venture between British Aerospace, British Telecom International, and GEC–Marconi, was the preferred vendor by the government to acquire and operate the satellite at 31 degrees west. Unisat initially proposed to offer Eurostar satellites for the UK DBS; they could carry two high-powered transponders plus up to four telecommunications channels for business services (26-watt TWTAs) or be configured to carry three high-powered channels. The three satellite systems, each with three transponders, were estimated originally by Unisat to cost $95 million per year, but under pressure from the consortium, the figure was reduced by 50%.

An alternative satellite system was proposed by Britsat, a private group intending to buy RCA satellites from the United States. They intended to offer a five-channel two-satellite system with the second satellite providing full back-up services. They claimed that the system would use 200-watt TWTAs, could be launched by 1987, and would cost $47 million per annum.

By June 1985 the whole project was in tatters. The DBS consortium, which had become known as the '21 Club' because of the number of

Table 5.5   Project: UK DBS (August 1985)

Responsibility for DBS	Home Office	Fixed
System owners	Consortium of BBC (50%); ITV (30%); Granada, Pearson, Virgin, Thorn EMI, Consolidated Satellite Broadcasting (all share 20%)	July decision indicates a no-go decision by the consortium
Satellite contract	Unisat	Unisat closed down. Britsat had offered an alternative system and was still trying to get support
Satellite system	Eurostar satellites, each $2 \times 230\,\text{W} + 5 \times 26\,\text{W}$ or $3 \times 230\,\text{W}$	Two satellite system proposed
Transmission standard	Modified C-MAC	No agreement with other European countries
Proposed services	Three channels: Film/children's News/live events Family entertainment	Never completely agreed
Launcher	Shuttle and Ariane reservations	
Launch date	1988 or 1989 proposed	Completely uncertain

organizations involved, decided that there was no future in the business and notified the Home Secretary accordingly. United Satellites subsequently closed offices. The various broadcast companies began to look at other options, such as a UK super channel. In July 1985 the Government asked the IBA to explore ways of putting together another DBS consortium.

In summary, the structure of the project may be represented as shown in Table 5.5.

5.9.3 FRANCE

Under the Franco-German industrial agreement, France commissioned Eurosatellite to build its first DBS satellite to be managed by Télédiffusion de France (TDF), the public body responsible for television and radio. This satellite, TDF-1, and a similar satellite being constructed for Germany, TV-SAT, are the first in a series of three for each country, all to be located at 19 degrees west (Fig. 5.3). TDF and TV-SAT are identically configured with 5×230-watt transponders, four of which can be operated simultaneously. (This was three until technological developments in 1984 made it possible to operate four.[20])

Until early 1984, this project had been the front-runner in Europe, with an expected launch (by Ariane) in late 1985. In mid 1984, however, the project was surrounded by controversy, as the French began reassessing the cost, emphasis, and development of 'new media', in particular, television, cable, and satellite. A further uncertainty with the project concerns the manufacture of the travelling wave tube amplifiers (TWTAs). AEG and Thomson–CSF are building six TWTAs each, three for TV-SAT and three for TDF-1. Thomson–CSF announced in November 1984 that there would be a 6- to 8-month delay in the manufacture and it is reported that AEG may also be having problems; in AEG's case the problem was with the power units.[21] The launch date is now late 1986.

Support for the project and the future high-power DBS comes from the Communications Minister, who was responsible for TDF. Opposition to the project has been led by the PTT, who also opposed the cabling of France using coaxial or copper wire cables (rather than optical fibre). Supported by the Direction Général des Télécommunications (DGT), the telecommunications area of the PTT, they have recommended that any follow-up satellites be of much lower power, allowing the satellites to offer a greater number of channels and, hence, be more economical. At this stage, however, both the French and the German governments have placed letters of intent with Eurosatellite for a second satellite in each series (TDF and TV-SAT), and an in-orbit spare. TDF-2 is tentatively scheduled for launch in February 1988 and will be a back-up satellite, providing one further transponder for the system.

A critical missing element in the project prior to 1985 was any market

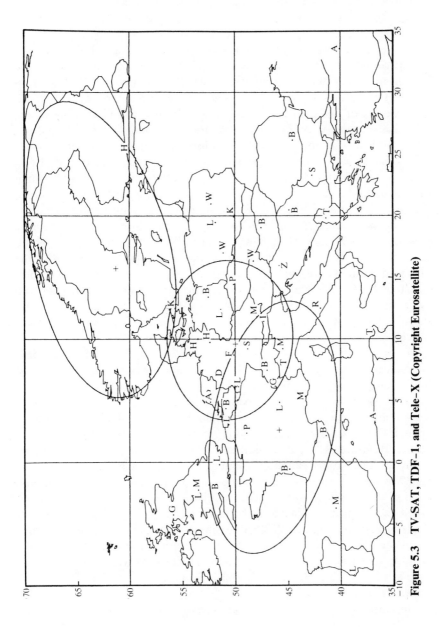

Figure 5.3 TV-SAT, TDF-1, and Tele-X (Copyright Eurosatellite)

109

evaluation of what services to provide and what to charge transponder lessees. It was not until August/September 1984 that French programming groups were associated with the project at all. In December 1984, a DBS marketing company, Société de Commercialisation des Satéllites de Télédiffusion, later to become Télévision par Satéllites, was established to manage these systems and to lease the transponders. The original proposal was that two channels would be leased by Compagnie Luxembourgeoise de Télévision (CLT), the national broadcaster in Luxembourg. Both governments (France and Luxembourg) signed an accord whereby CLT would have the use of two of the channels to offer French and German language advertiser-supported entertainment programmes: one, Radio Télé-Luxembourg (RTL), was to be in collaboration with French partners (a 40% equity share going to French-speaking radio stations such as Europe 1 and RMC of Monaco) and one, RTL-PLUS, in collaboration with Bertelsman (also a minority 40% equity share) for a German entertainment channel.

The understanding was that RTL would be the only French language advertiser-supported channel on French DBS—a real bonus. The strategic positioning by RTL with TDF was in response to the threat both organizations felt from the Luxembourg government's support during 1983-1984 of a medium-powered alternative project, Coronet, a project designed outside the WARC '77 plan (discussed in the next section). CLT had been awarded Luxembourg DBS fequencies by the Government, but the restricted footprint and power-flux-density limitations made the project uneconomic.

In early 1985, the position changed somewhat. The French President announced a plan to introduce private regional television networks, a first in French history (see Chapter 2). This undermined, at least in spirit, the privileged position that the CLT had negotiated.

In addition, the TDF marketing emphasis changed. The TDF footprint offered European coverage and this became the focus of the proposed service. By March 1985, Télévision par Satéllites was promoting this opportunity for a pan-European multi-channel, multi-language entertainment system. Most major programming groups in Europe were approached and an attempt was made to put together a private programming consortium to invest in the project. The French government, with a 33% stake, is the single largest shareholder. A notable omission is RTL. With the transmission system D2-MAC, the system adopted by the French government, it is possible to transmit four audio channels with each video channel. This enables French, English, German and Italian, or any four languages, to be transmitted simultaneously.

The turmoil in telecommunication policy in France in 1985 created considerable uncertainty for the project. By December 1985 three channels had been allocated. One is to the Berlusconi-Seydoux consortium for a European entertainment channel. One is to the UK publisher Robert Maxwell, who has been given exclusive rights to provide English language services on TDF. The third is to the French cultural channel organized by Pierre Desgroupes. It is an associate company of FR3 which will hold 45% of its shares.

DBS is expensive. To overcome what could have been an impasse in the negotiations with channel providers, the French have reduced the proposed annual rental for a transponder by 50% to around $8 to $10 million. This was achieved by:

1. Extending the time period to recoup technical costs of TDF from 7 to 10 years to 15 years
2. Increasing the number of channels against which costs will be allocated from 3 to 4
3. Absorbing (writing off) development costs, estimated to be $80 million, associated with TDF-1

In summary, the structure of the project may be represented as shown in Table 5.6.

Table 5.6 Project: French DBS, December 1985

Responsibility for DBS	TDF (technical) Télévision par Satéllites (marketing)	Fixed
System owners	TDF	
Satellite contract	Eurosatellite	
Satellite system	Up to 4 × 230-W with operational transponders	
Transmission standard	PAL	Long-term objectives
	D2-MAC	D2-MAC
Proposed services/lessee	Berlusconi-Seydoux	
	French cultural, Canal 1	
	Maxwell-Mirrorvision	Unallocated
Launcher	Ariane 2	
Launch date	Mid–late 1986	

5.9.4 WEST GERMANY

The TV-SAT project is the other half of the Franco-German collaboration in DBS (Fig. 5.3). The satellite, which is under construction by the Euro-satellite consortium, is identically configured to TDF-1. Each transponder will be 230 watts but there is a limitation of four simultaneously active channels on the spacecraft, although there is discussion of extending this to five channels with full redundancy. The satellite is likely to be launched after TDF-1, on Ariane 2, although scheduled to be launched before it, probably in late 1986 or else early in 1987. As with TDF-1, the TWTs are being developed by Thomson and AEG Telefunken. While the German company has not taken the same design approach as Thomson, it too has fallen behind schedule in manufacturing the high-powered tubes.[21]

The Deutsche Bundespost (DBP) is responsible for managing the satellite system, although like other PTTs it is not responsible for the funding of the project, and the political decision on the *use* of the transponders falls to the 11 Länder governments. They have not agreed on a unified media policy for West Germany and have not decided on the allocation of the five transponders, but have called for proposals from ARD, ZDF, private consortia, or mixtures of public/private consortia for the channels. The Länder governments have also agreed on some of the basic principles for advertising in their 'countries', and lessees will be allowed to advertise for up to 20% of the transmission time. There are restrictions on when the advertisements can occur, e.g., no programme breaks unless the programme exceeds 1 hour.

Table 5.7 Project: German DBS, August 1985

Responsibility for DBS	DBP (satellite)
	Länder governments
	  (transponder
	  allocations)	

System owners	DBP	
Satellite contract	Eurosatellite	

Satellite system	Up to 4 × 230-W
	  operational
	  transponders	

Transmission standard	D2-MAC	Proposed

Proposed services	Under review
	Could include ARD Eins
	  Plus, 3 SAT, SAT 1, and
	  a private pay channel
		Tenders have been
		  called

Launcher	Ariane 2	

Launch date	Mid 1986 to early 1987
	  although scheduled to be
	  launched before TDF	

5.9.5 SWEDEN

The Tele-X project is predominantly Swedish funded through the Swedish Space Corporation (Fig. 5.3). It arose as an interim experimental measure because of the failure of Nordic countries to agree on Nordsat, the proposed pan-Nordic satellite. A committee set up by the four Nordic countries involved in DBS, Sweden, Norway, Finland, and Iceland, has recommended a go-ahead for Nordsat and a merging of the two projects is still a possibility.

The satellite contract has been let to Eurosatellite with Swedish support; L. M. Ericsson and SAAB-Space are responsible for the satellite payload, the ground stations, and the TTC (telemetry, tracking, and command). Delays experienced by TV-SAT and TDF-1 apply to Tele-X as it is the same consortium making the travelling wave tube amplifiers (TWTAs).

The proposed satellite, a third in the TDF-1/TV-SAT series, has a very different payload. There will be five operational transponders: three high-powered 230-watt TWTAs for DBS services, of which only two can operate when the other 2×100-watt TWTAs are being used for data/video business telecommunication services.[22] The satellite system is to be managed by Notelsat, a joint venture of the Swedish and Norwegian PTTs.

The satellite is experimental and may form part of or be absorbed in the Nordic system. It is estimated to cost $110 million and is being structured to fit into the national and international telecommunications network. Its orbital allocation is 5 degrees east. The project has been plagued by dispute since its inception, the major issue of disputation being the appropriate allocation of costs across the respective countries.

The 'open-sky' policy for satellite reception in Sweden, a major liberalization in Swedish broadcasting policy, will facilitate market development of the services.

In summary, the structure of the project may be represented as shown in Table 5.8.

Table 5.8 Project: Swedish 'independent' DBS-Tele-X, December 1985

Responsibility for DBS	Department of Industry
System owners/managers	Notelsat
System contract	Eurosatellite
Satellite system	2×230 W (operational) $+ 2 \times 100$ W (for proposed data/video business)
Transmission standard	C-MAC
Proposed services/lessees	Film channel
	General entertainment channel
Launcher	Ariane 2
Launch date	1988 $(+)$

113

5.9.6 IRELAND

The Irish government put its DBS project into the market-place and invited tenders from individual companies and consortia around the world interested in funding and operating a DBS service. Access to the British market, given that the United Kingdom and Ireland share the same orbital slot of 31 degrees west, was the major incentive for would-be investors. The quid pro quo for the government was the perceived economic benefits (investment, employment, and new industry) for Ireland.

Four formal proposals were received by the Department of Communication when applications closed on 31 July 1984, and a decision was given in September 1985. The position in the United Kingdom does not give much comfort for a nation of 1 million households and a decidedly non-pan-European DBS footprint.

There is only one non-Irish-led bid and that is from United Satellites Ltd (Unisat), a United Kingdom consortium. The other three are: (1) Westsat, comprising RTE, Allied Irish Investment Bank, Guinness Peat, and An Bord Telecom Eireann; (2) Atlantic Satellite; and (3) the Irish Independent Newspaper Group, together with the Ulster Investment Bank. Strategies differ among contenders but in all cases one or two of the five allocated channels would ultimately be for the state broadcasting corporation RTE. The Independent Newspaper Group see the United Kingdom as a potential market once the UK DBS satellite is operational. They do not intend to launch until 1 or 2 years after that satellite.[23]

On the other hand, Atlantic Satellite and Westsat were the two most serious contenders. Atlantic has been reasonably aggressive, and believes that, with an American satellite system—the satellites to be provided by Hughes Aircraft Corporation through Hughes Communications Galaxy—a 1988 launch is possible. With a high-power/low-power configuration of transponders, they propose to offer telecommunications services, as well as relay services, across the Atlantic, in addition to DBS. Westsat also proposed a mixed television and telecommunications satellite, but had no definite plans for the satellite system itself.

The Government selected the Atlantic Satellite proposal. Their first step is to demonstrate economic feasibility of the project by mid-1986.

5.9.7 SWITZERLAND

The Swiss market is too small for a national DBS service to be economical. The orbital location of 19 degrees west coincides with France, Germany, and Olympus, and any Swiss service aims to capitalize on the fact that all European DBS antennas will be turned in that direction.

Competing companies lobbying to use the five DBS channels (Telsat, Teleclub, Universum Presse Njler of Geneva, and Peter Keoppeli—a private individual) have developed proposals to offer channels in German, French, and Italian languages, directing them (in terms of programmes and advertising) at the five-nation region—Italy, Austria, Switzerland, Germany, and France.

No firm decision on the long-term future of DBS has been undertaken by the government. All of the above proposals have been rejected and the government has taken a wait-and-see attitude towards the whole DBS issue.

5.10 Costs of DBS systems

The cost of high-powered satellites raises doubts as to the economic viability of systems with only two or three transponders. The equation is even more horrendous if the research and development costs for high-powered DBS are amortized against a system of two or even three satellites.

The Macintosh DBS report (1981)[24] summarized the 10-year investment implications for a DBS system such as TDF or Unisat. Drawing on their analysis and recent cost figures,[25, 26] we have the following position for a three-channel system:

Cost components	$ million
3 × satellites at $60 million (includes some R & D costs)	180
2 × launches at $36 million	72
2 × insurance at $30 million	60
Ground control installations	20
Total system investment (conservatively calculated)	332

The annual cost of a transponder lease depends upon the cost of borrowing money, the number of operational transponders, the annual operating costs, and the length of life of the system. Assume an annual operating cost of $4 million and the cost of borrowing money to be 15%. Then, excluding any profit in the equation, the annual cost per transponder becomes:

Number of operational transponders (on both satellites)	Annual cost ($ million) for project life of	
	8 years	10 years
4	17.1	15.4
6	11.4	10.3
8	8.5	7.7

As an indication of proposed costs the French are quoting an annual rental charge of $10 million per channel and the satellite costs alone for the three-channel Unisat system were estimated to be around $12 to $14 million per annum. Thus, programme providers are finding that DBS is not a cheap path of entry to their markets.

The total power payload of a satellite may be redistributed in a number of ways, for example, 700 watts may be either 3×230-watt channels or 10×70-watt channels. With a total system gain of around 10 dB since WARC, due to threshold extension and modulation gain, a 50-watt TWTA may provide an acceptable signal—not only nationally, but also for much of Western Europe.

By way of comparison consider a hypothetical medium-power satellite system (say 14×50-watt transponders). In this case there are no significant additional R & D costs to be allocated and the cost components become:

Cost components	$ million
2 × satellites at $35 million	70
1 × launch (booster and integrators included)	40
1 × insurance	12
Uplink and ground control station	20
	142

As can be seen, the significant cost savings arise when the total costs are allocated across 14 channels (or even 28 channels when there are two operational satellites) rather than three or five channels. For a system with 14 operational transponders and an annual operating cost of $4 million, the annual lease to cover costs becomes $2.25 million over 8 years or $2.04 over 10 years. If there are 25 operational transponders on a two-satellite system, the figures become $1.36 and $1.14 million respectively.

Although cost is important, an interrelated factor is the satellite coverage, i.e., the potential market for the service. If the national footprints are taken as given, it can be shown that greatly reduced TWTA radiofrequency power output will still provide the desired DBS service.

The trade-off between antenna size and respective satellite coverage in Europe can be demonstrated. Consider alternative satellite footprints across Europe. By estimating the number of households contained within each beam a comparison can be made among a high-powered satellite (TDF-1), a medium-powered satellite (the GDL proposed system), and a low-powered satellite (Eutelsat 1-F1) (Table 5.9).

The relative costs of a DBS system are finally beginning to become apparent to aspirant national bodies. No matter what reasons nations had for entering this field—nationalistic/altruistic/political—the bottom line is all important.

Table 5.9 Footprint comparisons of satellites in Europe

	TDF–1*		GDL/SES†	Eutelsat 1–F1
Transponder power (W)	230		45	20
Effective isotropic radiated power (EIRP) (dBW)	59.5	52.4	50.0	40.8
Comparable antenna diameter (m)	0.33	0.76	1	2.9
Household coverage (millions)				
Scandinavia	00.0	3.9	5.0	0.0
Mid-Europe	43.5	57.5	56.5	46.0
United Kingdom and Eire	6.5	21.7	19.7	20.6
Italy and Spain	16.8	28.1	19.1	00.0
Total (millions)	66.8	111.2	100.3	66.6
Percentage of all households	58	96	86	58

*The main TDF footprint with an EIRP of 59.5 dBW encompasses 67 million households. This would require a smaller receive antenna than the footprint of the medium-power satellite GDL, but as can be seen the picture has changed significantly from the WARC '77 time. In the case of Eutelsat 1-F1, the EIRP is too low for small receive antennas across all of its footprint.
† Derived from the SES (Société Européené des Satéllite) footprint.

5.11 Potential demand for DBS

The other side of the cost equation is the demand or market potential for DBS. We have suggested that most systems are still a number of years from being launched, and some may never secure the initial capital. Long-range forecasting of new media is notoriously unreliable. We believe that it is more useful to make a set of explicit assumptions and address the question from a 'what-if' perspective. In this way, the model may be modified over time as the assumptions change.

In a simplistic sense, the potential market for DBS is every television household which is not on a coaxial cable antenna system or a single unit master antenna system (MATV), but which falls within the satellite footprint. (The cable and MATV markets are discussed in Chapter 6.)

To give more substance to this market opportunity, we have considered the position in 1990 for a European DBS satellite. The general information is derived from TDF-1, but could easily be modified to apply equally well to the Irish, United Kingdom or German satellites or a medium-powered alternative such as the GDL project.

The objective is to determine if, under a set of 'reasonable' conditions, a DBS satellite with a pan-European footprint is an economic venture. The basic model assumes that if cable is available, a household will take it in preference to DBS based on cost and programming choice factors. Thus, all households cabled (and passed by cable) are excluded. This is not to say that a cable operator will not take the services, or at least some of the services on a DBS satellite for distribution to cable households. In any case, most countries with new media laws have a must-carry provision for locally originated DBS channels.

We assume further that the existing satellite master antenna television (SMATV) market will either be cabled, turn to DBS, or stay as a relay service for national broadcasters. Those MATV systems which do not take an Eutelsat or Intelsat service have the same chance of subscribing to DBS as other households.

Finally, it is assumed that the WARC '77 satellite footprint for TDF is still applicable, although modified to account for technological developments since 1977: i.e., to allow for smaller antennas than originally planned.

Under the above assumptions, the first step has been to estimate the number of potential DBS households across European countries, irrespective of a particular satellite system. This calculation is simply total households adjusted for cable and SMATV (Table 5.10).

To estimate the possible penetration for TDF, a probability model is used in which the probability that a household takes the hypothetical DBS service is assumed to be a function of:

1. The antenna size necessary to get adequate reception. This is a purely technological constraint linked to transmission power of the satellite and geographical location of the household. It is assumed that if the antenna is 0.30 cm in diameter or less, it is possible to erect it on 95% of the households. If the antenna is 1.2 m (or larger), only 10% of the households are able to erect such an antenna.

2. The attractiveness (utility) of the package of services offered on the satellite. A linear utility measure ranging from 0, when there are no services, to 1 for a DBS package of five services is assumed. The decision to take the service will be influenced by the range of other services on satellites at the same orbital location which can be received with the same hardware.

118

Table 5.10 Potential market for DBS, 1990*

Country	Households	Households (millions) Cabled or passed by cable	SMATV potential†	DBS potential	Estimated DBS penetration
Austria	2.5	1.4	0.6	0.5	0.07
Belgium	3.4	3.0	0.1	0.3	0.06
Denmark	2.0	1.1	0.6	0.3	0.03
Finland	1.7	1.0	0.4	0.3	0.02
France	17.9	4.4	6.3	7.2	1.77
Ireland	0.8	0.4	—	0.5	0.05
Italy	17.6	—	5.8	11.8	1.28
Netherlands	4.7	3.6	0.8	0.3	0.04
Norway	1.4	0.7	0.25	0.4	0.02
Spain	9.4	—	2.5	6.9	0.70
Sweden	3.2	1.2	1.0	1.0	0.06
Switzerland	2.2	1.4	0.65	0.15	0.04
United Kingdom	20.1	7.1	1.5	11.5	0.97
West Germany	24.6	8.5	7.5	8.6	1.64
Total	111.5	33.8	28.0	49.75	6.75

* Greece and Portugal are excluded from this analysis.
† From Table 6.6.

3. The price of the system. There are two components: first, the hardware price of the antenna, receiver, and decoder, and, second, the monthly subscription (if any) for the package of services. It is assumed that the equipment is priced from $300 to $500 across countries and that a monthly fee of $15 is imposed for the service.
4. The marketing effectiveness of the service. This includes marketing and promotion to ensure consumer awareness and also competitive positioning of the DBS service relative to competing DBS services and, where applicable, other discretionary income choices for the household, such as video cassettes. A non-linear ordinal scale was used where the various countries were allocated a value between 0 and 1, depending upon the existing level of VCRs and cable.

The probability model was applied under the above set of assumptions, having weighted each factor according to its importance. We also assumed a general European-wide entertainment package of services on the system. A two-stage process was used to estimate the market size. At the first stage, the probability of a household taking the service was computed, based on the technological assumptions of installing antennas of various sizes. At the second stage, the three factors, utility, price, and competitiveness, were weighted according to their perceived importance in the decision process for taking a DBS service:

Utility   40%
Cost   55%
Competition 5%

The outcome is a potential DBS household market for TDF in 1990 of 7 million households, a substantial opportunity (Table 5.10). The United Kingdom, France, West Germany, and Italy are absolutely and relatively the key potential markets for DBS. In addition, Spain, which is also un-cabled, is unlikely to be cabled, and is a nation with an unquenchable thirst for entertainment programming and could add 0.7 million households to the potential market for DBS.

5.12 Challenges to WARC '77—the future

There are a number of developments that are influencing the shape of DBS in Europe. It has been argued already that *technological* developments have changed the viability of a national focus for European DBS satellites.

5.12.1 DBS FROM COMMUNICATIONS SATELLITES

In exactly the same way that USCI in the United States used Anik C, a low-powered (20-watt) KU-band satellite, to show that it was technologi-cally feasible to provide early entry into DBS, it can be seen that Eutelsat 1–F1 can provide 'reasonable' pictures on 1.0 to 1.8 metre antennas over much of Europe. Telecom 1 and Intelsat V offer similar possibilities. This technical possibility is fast becoming an economic reality with the German and UK government decisions to consider legalizing downlinks from FSS satellites to the SMATV markets.

This array of cable relay services would also be available to those with home equipment and the necessary decoders, as well as necessary govern-ment licences. The inhibiting factors for using communications satellites are:

1. There is no uniform encryption system in Europe.
2. The ITU regulatory position on individual reception from telecommun-ication satellites is not clear. PTTs hold the position that receiving equip-ment for the FSS spectrum requires a PTT licence, whereas this is not true for the broadcast spectrum.
3. There is no agreed transmission standard for satellite-delivered television pictures.

At the end of 1985, the communications satellite choices for European viewers were:

1. (a) Intelsat VA (located at 27.5 degrees west).
      This footprint is centred over the United Kingdom and has four services currently being offered to UK cable systems:
      Mirrorvision
      ScreenSport/Lifestyle/Arts Channel (1986)
      Premiere/The Children's Channel
      Cable Network News
   (b) Intelsat V at 57 degrees east
      The Deutsche Bundespost leased six half-transponders on this satellite planned for launch in January 1985. The services are for general and regional programming. The German Music Box channel was the first lessee.

2. Eutelsat F-1 (located at 13 degrees west).
   There are nine operational transponders, providing entertainment programming (Table 5.11).[27]

Table 5.11 Eight transponders providing entertainment

Service	Originating country	Content
Sky Channel	United Kingdom	General entertainment
Music Box	United Kingdom	Pop videos, music, concerts
TV-5‡	France	Drama, music, current affairs, drama from national retreats
SAT-1	West Germany	General entertainment
3-SAT (ZDF, ORF, SRG)*	West Germany	Mixed programmes, drama from national networks
Téléclub	Switzerland	Films
RA1	Italy	Current affairs, opera, entertainment
Europa	Netherlands	News, entertainment
ATN/Filmnet	Belgium	Films, entertainment
RTL-Plus	Luxembourg West Germany	Entertainment

* This transponder is an East Spot beam. The rest are West Spot beams.
† To be changed in 1986.
‡ A Norwegian channel, New World Channel, operates on the TV-5 transponder when TV-5 is off the air.

5.12.2 MEDIUM-POWER ALTERNATIVES

The project which has caused most concern to the 'orderly' development of DBS and which was seen as an overt threat to Eutelsat was the proposed Grand Duchy of Luxembourg's Coronet project.[28]

Introduced initially as a cost-effective FSS alternative to Luxembourg's proposed DBS venture and as a possible means of establishing a support industry base in Luxembourg, the project is Europe's first 'breakaway' satellite system. The aim was to sell private investment in the satellite system, with the franchise to be held by Société Luxembourgeoise des Satéllites (SLS), while allowing the Luxembourg government to maintain a majority institutional control. In that sense, it was still a national telecommunications satellite system. The project had obvious appeal to the Luxembourg government. While CLT, the national television and radio broadcaster, is the largest contributor to tax revenue in the Grand Duchy (5 to 7% of total levies), it is not under Luxembourg government control. A second major shareholder in the project was Coronet Research Inc., a Luxembourg company headed by Clay Whitehead, formerly chief executive of Hughes Communications, who acted as a consultant and adviser to the Luxembourg government.

Core investors in the project were approached to raise $5 to $10 million as a downpayment for an order for two medium-powered satellites (16×50 watt) from an American supplier. The intention was to restrict equity to European companies and so minimize the likelihood of being seen to be an attempt by the United States to take over the European satellite business. Where possible, it was proposed to let the subcontracts to European companies. Transponder lessees who wanted to offer cable relay, MATV, DBS, or telecommunications services would then be sought. By early 1985, potential investors had begun to emerge. Beijer, a Swedish investment company, took an 80% equity stake and options on two transponders. Dutch and French corporations were also negotiating with Coronet for transponders and Home Box Office (HBO), USA, held a small equity position.[29]

Although the IFRB filing was lodged, the project was seen as a direct threat to the other European DBS projects, to Eutelsat, which is currently offering telecommunications satellites for cable relay and data/video services, to Intelsat, and to European PTTs, who wish to control satellite delivery of telecommunication products within Europe. The US proposals for trans-Atlantic transmissions of telecommunications traffic including cable relay services and even direct live transmissions from America (International Satellite Inc., RCA, Orion Satellite Inc., and Cygnus Satellite Corporation) gave further concern to European governments, in their capacity as members of Intelsat, to Eutelsat, and to the national PTTs. If cracks can be found in the PTT web, such as through Mercury or the privatization of

British Telecom in the United Kingdom, the whole market may become open to internal and external competition.

The change of government in Luxembourg in June 1984, renewed pressure by CLT and the French, the signing (but not formal ratification) of the French–Luxembourg government accord for TDF, threats by the Germans that downlinks would not be approved in Germany, and the need for Franco-German industrial support for Luxembourg clouded its future as proposed.

The project took a new direction in 1985. On 1 March 1985 the Société Européené des Satéllites (SES) was formed to operate the GDL satellite system. Coronet, HBO, and Beijer were no longer among the investors. The structure involves class A shareholders who have 80% of the equity and 66% of the voting power (Deutsche Bank, Compagnie Financière, Banque Générale du Luxembourg, Banque Internationale à Luxembourg, Dresdner Bank International, Kinnevik International, Kirkbi, Natinvest SA Luxembourg, a holding of private investors, Société Générale de Belgique and RITA, Réalisations et Investissements en Technologies Avancées) and class B shareholders each having 10% equity and between them 33% of the voting power (Caisse d'Epargne d' État du Grand-Duche de Luxembourg, Société National de Crédit et d'Investment).

The company, with initial shareholder capital of $5 million, began discussions with American and European satellite manufacturers. Its decision to proceed depended as much upon securing the additional funding, a decision made easier by the Luxembourg government's decision to offer a guarantee for at least part of the total satellite cost, as it did upon steering a path through the regulatory environment. In September 1985, SES announced its decision to enter the market by commissioning an RCA satellite for launch in 1987 and then in November announcing a decision to use Arianespace to launch the satellite.

Technologically, a medium-power system is a natural successor to the low-power systems. But the project is also subject to the economic harm clauses in the Eutelsat Agreement as the satellite is to operate in the telecommunications section of the frequency spectrum. The project does have support within Europe and it could offer new television channel providers an opportunity to enhance their markets by transmitting from the one satellite.

This project is the beginning. EBS (European Business Satellite), a Swedish-based project to compete with Tele-X, has announced its intention to launch two satellites in 1988 for private business communications. The satellites, which are to be off-the-shelf RCA or Hughes satellites, each with 14×20-watt transponders, give a European footprint not unlike the current ECS satellites, i.e., with a beam-edge EIRP of around 40 to 45 dBW.

The project, which is to cost $200 million, is at an early stage. Financial backing comes from Swedish venture capital companies led by Skandia Investments. Licences to operate the system have been applied for in the United Kingdom, Sweden, and Norway, but none have been granted as yet. This is considered to be a somewhat speculative venture.[30]

5.12.3 EUROPEAN MARKET FOR DBS—NON-NATIONAL MARKETS

The French plan, to operate TDF-1 as an economically viable service with a market limited only by the footprint and the attractiveness of the programming, is a step towards an open-skies policy within Europe. West Germany, the United Kingdom, and Ireland are all evaluating the size of their potential markets, not just their national markets, as they have all realized that national DBS using expensive high-power technology is not feasible—i.e., unless the respective governments support the projects financially.

Although there were no frequencies allocated for European services at the WARC meetings, the principle of a European television market is not new. It was embraced by the European Convention on Human Rights and Fundamental Freedoms (1950),[31] which outlined the right of freedom of expression:

> ... to hold opinions and to receive and impart information and ideas without interference by public authority and regardless of frontiers.

This was endorsed and elaborated in 1984 by the EEC which produced a Green Paper called *Television without Frontiers* in May 1984,[32] advocating that:

> ... cross-frontier broadcasting would make a significant contribution to European unification ... to help the peoples of Europe reorganize the common destiny they share in many areas.

Economic reality and regulatory 'push' are moving in the same direction. This is not to say that all of the DBS services will be pan-European. Economic markets rather than national markets will become the focus of attention.

5.12.4 COMPETITION FROM OTHER MEDIA

So far, the consumer has not spent $1.00 on DBS. Market forecasts indicate that from 10 to 20% of households may one day subscribe to appropriately priced and interesting services. In some countries, the extent of cable development already makes DBS a hard sell (Netherlands, Belgium, Switzerland) (see Chapter 6).

In the United Kingdom, the BBC and ITV companies have must-carry protection on cable for their broadcast services *and* for their proposed DBS service. So far, cable companies, where the service is being developed without government subsidy, are uncertain as to the economic viability of their proposed ventures. DBS could be a cost-effective alternative to cable if the dish size is small enough to satisfy local borough planning regulations and the sharp eyes of the various societies seeking to preserve traditional buildings and households.

Perhaps the real wildcards for most European countries are:

1. Will SMATV, using communications satellites, take the 'cream' off the potential market for DBS?
2. Will technological improvements in receive equipment mean that high-power DBS is obsolete before it starts?
3. Will the high penetration of video cassette recorders militate against or encourage the growth of the DBS market?

It would seem that VCR penetration may militate against pay services, but not necessarily against advertiser-supported or pay-per-view services. Its time-shift function may even be an advantage as DBS will certainly offer viewers more programme variety at any given time.

5.12.5 TECHNICAL UNCERTAINTIES

Two major technical uncertainties are associated with DBS, assuming that the design problems with the high-powered amplifiers have been solved.

First, European governments have not agreed on a common transmission standard, although there is general agreement to use DBS to introduce a better standard than either PAL or SECAM. At present, the United Kingdom, Norway, and Sweden support C-MAC, and the French and West German governments have announced their intentions to adopt the D2-MAC packet transmission system and to use it with their respective DBS projects. Already the chip manufacturers are trying to develop a compromise chip set which can adopt both transmission standards with only minor modifications to the two systems.

MAC (multiplexed analogue components) is a fine division multiplex system proposed by the IBA and recommended to the UK government by the advisory panel on technical transmission standards. The television signal is broken down so that the luminance and chrominance are carried out at different times. Thus to transmit both luminance and chrominance in one television line, the signal must be compressed in time. This is achieved by using a proportionally wider bandwidth. The signal is frequency modulated (FM) by the satellite and fits within the standard 27-MHz channel bandwidth allowed for under WARC '77.

It offers improved picture quality over the current PAL (phase alternation line) system used in terrestrial broadcasting in much of Europe, can be used as a bridge for digital transmission, is well suited for scrambling, and offers increased audio and sound capacity on a DBS channel. Calculations post-WARC show that these improvements, when transformed into reduced antenna sizes, provide a very promising picture (Table 5.12).[33]

Table 5.12 Comparison of PAL and MAC

Transponder power output (W)	Antenna diameter at 3 dB contour (m)	
	PAL (1986)	MAC (1987)
20	1.4	1.0
50	0.9	0.6
100	0.6	0.45
230	0.4	0.25

The digital sound and data are organized into packets and transmitted on a phased-shift keying system at 3 Mbit/s. Up to eight such high-quality sound services may be carried with one television picture, or a lesser number if data capacity is required.

It is the method of sound transmission associated with MAC that has led to the major debates in Europe. The EBU proposed three methods of sound transmission (A, B, and C) and C-MAC, a system which has the greatest capacity—it offers up to eight sound channels—by switching the complete transmission channel into digital mode during the intervals between picture lines, has been proposed by the UK government.

C-MAC has not been well received by the other European nations, even though it was put forward by the EBU Administrative Council in July 1983 as a recommendation to be used as a common standard for DBS. The French and German opposition was due to the inclusion of 3 Mbit/s of data in the video stream which means that C-MAC in its cable form uses 10.5 MHz—too much bandwidth for existing cable systems in France and West Germany, which have a channel bandwidth of only 8 and 7 MHz respectively. Two organizations, Thomson and Philips, designed a further MAC variant, called D2–MAC (the D2 meaning divide the data rate by 2), which carries data at a slower rate, 1.5 Mbit/s, and in effect reduces the number of sound channels from eight to four per video channel. Some compression of the video then gets the signal into a 7-MHz bandwidth.

By late 1984, the British and French seemed prepared to compromise. A solution, the '20/10' was proposed, whereby C-MAC and D2-MAC would both be modified so that the two standards would be compatible on a single set of decoder chips. The 20/10 refers to the two different data rates proposed by the British (20 MHz) and the French (10 MHz).[18] However, the official French policy to move straight to D2-MAC and further delays in the UK project indicate that the D2-MAC may emerge as the practical standard for Europe.

The second unresolved area has to do with encryption. Agreement does not exist, even at the national level, but while this is not a problem for the satellite provider, a consequence of this uncertainty is that European receiver hardware manufacturers have been slow in developing equipment for mass markets. The viability of a DBS service is enhanced significantly if the viewer can pick up all of the channels at a chosen orbital location.

5.13 Conclusion

Satellite transmission is shaping European cable and is about to impact the household directly first through SMATV and then through individual reception, but not in the grandiose way considered at WARC '77. The extensive use of communications satellites for cable relay services has saved the Eutelsat programme from being a financial embarrassment to the European PTTs. In the longer term, the television services may gravitate to higher-powered Eutelsat satellites or to the national DBS satellites, but only if the footprints offer economic markets given the expenses involved. However, technological advancements in the design of earth station equipment questions the need for the high-powered satellites as originally conceived.

The WARC picture no longer applies in Europe, at least for reception. The nations looking to DBS as a national industry, no matter how favorably, are finding that the costs and uncertainties are too great to try and adopt and then implement the WARC plan.

A number of factors have been identified which are necessary for European DBS to evolve as an economically viable service. The following is not definitive, but indicative of the major obstacles still remaining:

Technical
—An agreed transmission standard at least among countries at the same orbital location.
—The correct functioning in space of the as yet un-space-tested high-power transponders.

—Availability of receiver equipment at mass market prices. This assumes, implicitly, that the transmission standard is resolved and encryption systems or system are agreed.

Political
—National agreement to the concepts of television transmission across national borders.
—Favourable decisions on licensing of household receiver equipment.
—Local government planning permission for antenna installation forthcoming.

Economic
—Viability of high-power DBS either through government support or market pays.
—Acceptability of pay-television for European households, most of whom already pay a television licence fee.
—Evolution of pan-European advertising.

Social
—Acceptability of a three- or four-channel television service to householders.

Administrative
—National and local equipment and installation and sales forces in place to sell and install the service.
—National and possibly multi-national collection systems, if a subscription service is offered.

The DBS position is not promising. National approaches to DBS are different, the projects are structured differently, the reasons for entering the field are many and varied, and the challenges have already been mounted. So far the national governments have bought time by ensuring delays in private sector challenges, such as Coronet, but they have also experienced delays on their own projects.

The government commitment, at least in France and West Germany, ensures that the first round of high-powered satellites will be launched. However, technological and market uncertainties have resulted in delays of 2 to 3 years from the original 1984 proposed launch dates. The SES medium-powered satellite is scheduled for launch by May 1987, thereby ensuring that programmers have a variety of options.

Already, market forces, consumer demand, and economic viability are beginning to influence the direction of future satellites in Europe and question the economic wisdom of high-power satellites with minimal numbers of transponders.

References

1. C. L. Cuccia, 'Satellite technology matures in Europe', *MSN*, March 1979.
2. European Space Agency, *ESA Charter*, European Space Agency, Paris, September 1982.
3. European Space Agency, *General Information on Ariane*, European Space Agency, Paris, December 1981.
4. Arianespace, *Ariane V10*, Arianespace, France, August 1984.
5. S. Baker, 'In charge of the Eurobeams', *Cable and Satellite Europe*, March 1984.
6. European Space Agency, *ECS Data Book*, Technical Appendix, European Space Agency, Paris, 1983.
7. Eutelsat, general information, Paris, Autumn 1984.
8. P. Bartholome, 'Satellite broadcasting in Europe', *ESA Bulletin*, May 1984, pp. 6–11.
9. 'Another trans-Atlantic satellite service proposed', *Broadcasting*, 22 August 1983, pp. 39–41.
10. 'RCA Americom also to modify SATCOM 6 for European beams', *Satellite Week*, 20 February 1984, pp. 1–2.
11. Mark Oderman and Susan Crowe, 'The meeting of domestic and foreign policy', *Satellite Communications*, January 1984, pp. 22–4.
12. 'Reagan: separate international satellites in "national interest"', *Satellite Week*, 3 December 1984, pp. 1–3.
13. British Aerospace, 'ECS-1 technical information', British Aerospace, Stevenage, England, 1983, p. 3.
14. Eurosatellite GmbH, 'Eurosatellite', Munich, West Germany, 1983.
15. Commission of the European Communities, *Realities and Tendencies in European Television: Perspectives and Opinions*, Commission of the European Communities, Brussels, 1983.
16. Home Office and Department of Industry, *Direct Broadcasting by Satellite—Report of the Advisory Panel on Technical Transmission Standards*, Her Majesty's Stationery Office, London, 1981.
17. Home Office, *Direct Broadcast by Satellite*, Her Majesty's Stationery Office, London, 1981.
18. 'C-MAC: the 20/10 solution', *New Media Markets*, 11 December 1984, p. 5.
19. 'Unisat price too high', *Cable and Satellite News*, January 1985, p. 40.
20. N. L. H. Cresdee, 'A review of French satellite TV prospects', *Satellite TV News*, December 1983.
21. 'Travelling wave tubes among problems that will delay German and French DBS', *DBS News*, 11 November 1983, p. 6.
22. Jean Germain, 'Tele-X: a multipurpose communication satellite for the Nordic countries', *Space Communications and Broadcasting*, September 1984, pp. 229–35.
23. 'Irish DBS: four groups want in', *Broadcast*, 17 August 1984, p. 39.
24. Mackintosh International, *Satellite Broadcasting*, Mackintosh International, London, 1981.
25. 'DBS UK', *New Media Markets*, December 1984, pp. 2–5.
26. 'Satellites', *New Media Markets*, January 1985, pp. 17–18.
27. 'European satellite TV stations update March 1985', *Interspace*, 29 March 1985, pp. 8–9.
28. 'A Coronet for the Grand Duchy', *Cable and Satellite*, July 1984, pp. 10–12.
29. 'Coronet satellite project wins backers', *Ogilvy and Mather Euromedia*, January 1985, paragraph 8.

30. 'The Swedish satellite', *New Media Markets*, July 1984, pp. 4–5.
31. European Convention for the Protection of Human Rights & Fundamental Freedoms, Article 10, Rome 1950.
32. Commission of the European Communities, *Television without Frontiers*, Commission of the European Communities, Brussels, 1984, p. 28.
33. Patrick Cox, 'Is there going to be an "open skies" policy in Europe?', Proceedings of 'Cable television and satellite broadcasting', *Financial Times*, London, 19 and 20 March 1985.

6

Cable television

6.1 Background

Cable television evolved simultaneously with broadcast television during the fifties in both Europe and the United States. Nearly every European country introduced some form of cable television during this 25-year period and the systems fell under the jurisdiction of the PTTs.

On both continents, the reasons for introducing cable were the same: to provide a higher quality of television reception. The early areas to be cabled were either remote communities which could not receive signals from the network of transmitters or communities where reception was of a poor standard. In the United States, entrepreneurs erected private receivers to capture *existing* broadcast signals from adjacent communities. These signals were then distributed to individual households by cable. These were the original cable antenna television systems (CATV), systems that refer to groups of cabled homes. Provided that there were no new services included in the system, no Federal Communications Commission licence was required to simply retransmit existing broadcast signals to households, and the operator could charge a fee for this service. By contrast, in Europe, the PTTs operated the systems as a public service, generally without any profit inventives.

Many European cities also suffered poor reception due to signal interference from other broadcasters. Master antenna television (MATV) systems were developed to improve transmission to single-site areas, such as apartment complexes or blocks of houses. MATV and CATV systems in Europe are often indistinguishable, which results in a wide variance of published cable penetration levels.

In the United States, cable franchises are granted throughout the country on a local level. Once a local community has determined that it wants to have a cable system constructed, a request for proposal (RFP) is published, detailing the specifications of the community. Potential franchise operators then make their bids, and one operator is selected and granted a monopoly to operate in that area. The local telecommunications service is responsible for the laying of cable, but is precluded from being the cable operator, and the entire process takes about 2 years.

Between 1978 and 1982, the number of cabled households in the United States doubled to over 28 million. During 1981, 1982, and 1983, connections increased at annual rates of 16.9, 20.9, and 26.5% respectively.[1] In 1980, the FCC further deregulated cable services which allowed syndicated services to be set up in competition with local independent channels.

Cable networking, as such, did not start in the United States until September 1975, when Home Box Office, Time Inc.'s pay-movie channel, began to offer its service to cable operators by way of Westar satellite. So began a cable programming explosion over 9 years which resulted in 40 new channels (not all of which have survived). The programme services that were introduced earliest experienced the most success in terms of growth and subscribers. The one exception, Music Television (MTV), began in the autumn of 1981, but was the first programme service of its type and filled a gap in the cable programming market. The services offered in 1985 are broadly grouped into types in Table 6.1.

Table 6.1 Services in the United States in 1985[2]

Service type	Number of services
Pay-movie	5
Entertainment (including superstations)	7
Special interest	28
News/public affairs/ information/business	9
Religious	6
Music	4
Other	9

6.1.1 CABLE DEVELOPMENT IN EUROPE

In Europe, cable progressed more slowly than in the United States. Cable systems were developed specifically to retransmit the national channels and, in some central European countries, some foreign broadcasting. The most highly cabled countries, Belgium and the Netherlands, originally established their cable systems to operate a radio relay system. Switzerland, which is also highly cabled, developed its systems to eliminate interference from the mountains. In the remainder of the European countries, cable was used sparingly and primarily to improve transmission in specific areas.

The PTTs in Europe have historically had a monopoly over the cable systems (with the exception of Austria, Belgium, Luxembourg, and Switzerland), including the links, transmission equipment, facilities, and head ends. They construct and install the systems and are responsible for their maintenance. This monopoly is protected by a 1960 Council of Europe agreement, the *European Agreement on the Protection of Television Broadcasts*.[3] The agreement entitles states to authorize or prohibit CATV distribution within their borders. As a result, the monopolies have maintained strict control over cable systems and the cable industry has not become competitive as in the United States. In many systems, the PTTs perform customer relations and collect subscriber fees as well.

Typically, for a European community to become cabled, the local municipality must apply through the PTT and a majority of citizens must want the network. The PTTs then license operators who are responsible for running the systems on a local level. In most of the European countries, the municipality is, historically, either an active partner or the principal licensee of the cable system. Demands for a return on capital investment are therefore more flexible than for a commercial operator using borrowed money.

The relay of foreign terrestrial broadcast television signals on cable systems has been allowed in central Europe. Most advertising breaches between countries have been ignored as nations have tended to allow unintentional spillover, and intentional spillover signals, such as RTL into Belgium, have been legitimized. Most local systems are not permitted to produce their own programmes or broadcast local programmes that might compete with the national services. Prior to 1984, pay-television in Europe was not permitted except in Finland and Switzerland, and in France and the United Kingdom on an experimental basis. (See Chapter 8 for a discussion of the pay-television systems.)

The sudden growth of the cable industry in America encouraged European countries to examine the television industry in a new light. Extensive information and research on the US market was available, lending credence and support to the potential. Many would-be entrepreneurs felt they could build upon the US experience and establish similar systems for Europe. European governments, too, were looking for new ways to stimulate their declining industrial sectors.

The potential of using cable for the interactive services sparked further interest. Teletext and videotex services, which could be effectively distributed through cable systems, were more developed in Europe than in the United States. Home computer link-ups and interactive capabilities made cable of interest to retail firms, financial institutions, and computer software providers with ideas of establishing direct connections with consumers. (See Chapter 9.)

It was, however, the construction and launch of the ECS pan-European satellite Eutelsat 1–F1 that made a new European cable industry possible. Eutelsat 1–F1 provided the first outlet for cable operators to distribute a service to a large enough subscriber base to consider cable as a business. In addition, the PTTs relinquished a little control, as it was apparent that their monopolistic position—especially with television relay—was becoming increasingly difficult to protect. The inability to effectively stop foreign signals has led to a more flexible attitude towards cable than broadcasting. New regulatory guidelines reflect this changed attitude (Chapter 10).

6.2 Cable USA—1985

While Europe is only beginning to develop its 'private' cable industry, the steady growth in the United States has tapered off, indicating that there is a saturation level for audiences and an economic cut-off for cabling rural communities. The United States is now in its third phase of cable development, having completed its first phase of new cable systems initiation and the second phase of extensive building. The third phase emphasizes research and marketing with some upgrading and some new building. The industry is beginning to focus more upon increasing audience size within the operating systems; Europe is still in the second phase of growth and can learn much from some general trends of the cable industry in the United States today.

There are 5750 cable systems in operation in the United States, and over one-third do not have pay services. The largest number of systems fall within the 1000 to 4000 subscribers category, while the systems with 20 000 to 50 000 subscribers account for the greatest number of total subscribers.[4] Only less than 5% of the systems are two-way cable systems.

RCA Americom's SATCOM III, a low-power C-band (7 to 8 GHz) satellite, initially provided most satellite-delivered cable services to operators, carrying all major basic and pay channels in operation. Hughes Communications Inc. has recently created the 'Galaxy Club' on its satellite, Galaxy I, by selling transponders 'condominium' style so that the system comprises primarily those services that are the most successful.

The satellites with the largest number of services usually carry those services with higher ratings. While it is difficult to say which comes first, it is apparent that increased exposure through transmission via a 'popular' satellite produces more subscribers, and the more successful services induce operators to purchase the hardware (antenna, receiver, and low-noise converter (LNC)) to receive these services from their respective satellites.

The number of cable systems under construction is decreasing each year, and growth has slowed. Current prices for newly wired systems are around $20 000 per mile, while rebuilt systems cost about $3600 per mile. The cost to wire a single home is estimated to be $900 to $1200, with the average number of households per cabled mile ranging from 10 to 20 for new systems.[5]

Most of the companies involved in the cable channels are involved with other media as well, such as Warner Amex, Group W, ABC, Time Inc., and Turner Broadcasting. In addition, many of the programme services offered on cable are owned by these same companies. This is particularly true for pay-television, where Time and Viacom/Warner Amex control 90% of the US pay-television market.[6]

Advertiser-supported services have been increasing their viewership while pay services have declined slightly. During an average day, 62 to 65% of cable households tune into some advertiser-supported cable station for 6 minutes. During an average week, this figure increases to 82 to 90% and for an average 4 weeks, it further increases to over 95% of cable households tuned to an advertiser-supported channels for 6 minutes[7]—still not a major threat to the entrenched position of the networks. Cable advertising has not made a dent in the network television share of total advertising, and viability of advertiser-supported channels comes through a combination of advertising and a subscriber fee charge to cable operators to carry the service.

News is the most popular concept, although not the most profitable, for a cable channel, followed by films and music. Films also provide the strongest reason for subscribing to cable services.

Yet cable households provide a number of characteristics that are highly desirable to advertisers.

1. They watch more television during all day parts.
2. They have more viewers in the 18 to 49 year old age bracket.
3. They have more affluent households, i.e., a large percentage of audience with incomes + $30 000 per year.
4. Households of larger than average size.
5. Households with more than average number of children and teenagers.
6. Cable viewers more likely to purchase specific products, such as those related to new media (VCRs, video games, computers) and those associated with 'better living' (wine, suntan lotion, health care products).

However, advertisers still tend to use cable as a residual buy, knowing that reach and frequency (coverage and opportunity to see) are both best achieved through the networks. Cable research through the Cable Advertising Bureau (CAB) indicates that there is a need for emphasis on sales, service, and marketing, particularly at a local level. Most operators are technically oriented and need to be better informed about marketing their services. Local insertion equipment, i.e., electronic equipment which allows local cable operators to introduce their own advertisements in the transmitted programmes, is becoming popular with multiple-system owners (MSOs) and it offers one way for cable operators to tap into regional and national advertising markets.

By 1990, cable penetration is expected to rise to around 50 million house-holds from some 36 million households in 1984.[8]

6.3 Europe

The situation in Europe by mid 1985 was quite different from the United States. The European cable industry is approximately where the US industry was about 10 years ago. Cable television services have begun to be deregulated throughout Europe, allowing for the introduction of private and commercial services. At present, there are an estimated 12 million cable households in Europe.[9] Table 6.2 illustrates the size of the cable systems.

Table 6.2 Cable systems by number*

	Cabled households (millions)	Major CATV head ends
Austria	0.60	30
Belgium	2.80	80
Denmark	0.32	128
Finland	0.18	44
France	0.75	160
Ireland	0.24	3
Italy	—	—
Netherlands	2.58	550
Norway	0.26	39
Sweden	0.20	45
Switzerland	0.95	300
United Kingdom	1.95	83
West Germany	1.47	860
Total	12.30	2322

*Calculated from our own figures and data from Mackintosh International.[10]

Cable system expansion is predicted for nearly every European country, with the exception of Italy, Portugal, Greece, and Spain. Italy, with its private television systems in place has no need to expand programming through cable, and the other three countries are still improving their terrestrial services. In the remainder of Europe, most cable growth will be focused on six countries: Denmark, France, Sweden, West Germany, the United Kingdom, and Austria. Belgium, the Netherlands, and Switzerland are already highly cabled and will be discussed first. The national pay and advertiser-supported cable channels are described in Chapter 8.

6.3.1 BELGIUM

Belgium is the most highly cabled country in Europe, with over 80% penetration of all households. It was cabled in the sixties to establish a radio relay system. While little growth is predicted, the introduction of new services, rather than relays of terrestrial channels, is expected. In November 1984, a proposal to determine the future of cable programme services was put before the Flanders Council of State. It recommended that commercial television, satellite, and pay-television channels should be licensed within Belgium. Sky Channel, followed by Music Box, began their cable network penetration in September 1985. Both must supply a set amount of Belgian-made programming, Filmnet, the Dutch pay-TV channel, was also authorized to be received at the same time.

6.3.2 NETHERLANDS

The Netherlands has over 70% of the households linked to cable systems established as part of a radio relay system begun in the sixties. However, it uses old systems with limited channel capacity. The government has given priority to pay-television which has encouraged new cable development and upgrading of the existing systems by the large system operators. While the PTT operates many of the cable systems, small operators have begun to develop cable systems to expand the reputedly 'boring' domestic television programming under the technical specifications laid down by the PTT. The CAESMA, which is a semi-public company composed of the NOS and the PTT, operates 15% of the small systems. The government is in support of further expansion because more jobs would be provided.

The main source of finance is from subscriptions, and the Scientific Council, which has made recommendations as to the further progress of cable in the Netherlands, has recommended that the subscribers themselves should decide upon the channels they receive according to the area in which they live, which would include local channels with some advertising. As a result, channels distributed via satellite can now be received on the Netherlands cable systems. The advertising regulations for foreign signals differ from those for national broadcasters, being more liberal in some respects (Chapter 10).

The Netherlands has also planned broadband experiments in Zaltbommel, Dortrecht, Amsterdam, Geldiop, and Limburg. Limburg is the most advanced of the experiments with plans to provide 100 000 subscribers with an interactive fibre optic cable system.

6.3.3 SWITZERLAND

Switzerland is the third most cabled country after Belgium and the Netherlands. Its broadcasting organization, SRG, caters to the French-, German-, and Italian-speaking populations.

Pay-television is legal in Switzerland and a German-language channel, Teleclub, is available. A French-language channel, run by Telecine Romandie, began in December 1985. It is funded by Swiss shareholders, including SRG, the national broadcaster,[11] and aims to provide three categories of programmes: (1) movies; (2) an adult segment; and (3) a category which includes music, sport, and children's programming during the day and adult programming late at night.

Swiss cable has reacted quickly to the opportunities through Eutelsat 1-F1. It was one of the first countries to take the English-language entertainment service Sky Channel and most major cable operators have installed antennas to receive signals. The majority of these systems can take up to 20 channels, so there is little need to invest in new hardware. Although there are an estimated 1200 cable systems in Switzerland, most of which are small, the top 20 systems account for the majority of households.

Pay-television development, however, has been very slow. Penetration is low, averaging only 5% of potential subscribers by mid 1985, and is not very attractive for cable operators who must install set-top decoders (separate 'black-boxes' to unscramble the signal, usually placed on top of the television set) into subscriber households or introduce built-in switching into their systems. Autophon, a Swiss cable operator, is supplying decoders.[11] Autophon will also supply the Racal-Oak decoder to Telecine Romandie.

6.3.4 DENMARK

Over 50% of the viewers in Denmark have access to cable systems and, until now, Denmark has had little need to expand its systems. However, a need has been created to introduce foreign services, particularly from West Germany, Norway, and Sweden, to supplement the single government channel. A law of 15 June 1973 prohibits MATV and CATV systems from distributing any programmes except those from Radio Danmarks and foreign broadcast authorities. The Ministry for Cultural Affairs banned the distribution of Sky Channel in Denmark because it conflicts with the state monopoly. The government is anxious to develop a broadband cable system for the entire country, although work has not yet begun, and has authorized experimental distribution of local programmes in the future. In the meantime, the state has continued to ban the reception of satellite-distributed services.

6.3.5 FRANCE

While France is the least cabled major European country, the government has plans to develop a national broadband service with a fibre optic system. The November 1982 plan from the French government 'Plan Cable' calls for the installation and construction of a fibre optic network by the PTT. The legal structure finally agreed upon authorizes the PTT to lay, own, and maintain the system and charge cable operators 42 French francs per subscriber. The cable-operating companies are themselves 'mixed economy' organizations known as SEMs (Société d'Économie Mixte). They require local authorities to have a minimum of 33% equity participation and restrictions are planned to limit private corporate participation in more than one SEM.

The plan calls for a specific number of households to be cabled each year, with an optimistic forecast of 0.75 million households connected by 1986, giving France a 4% cable penetration and then, after 1986, the plan is to be accelerated so that 1 million households per year will be cabled through 1992.[12] During the first 2 years, coaxial cable is to be used to give the industry time to manufacture optical fibres. The fibre optic system is to be used together with the telephone systems to provide information services. The revenue generated from the information services is to be a major source of finance for the plan. There are to be 20 networks established with 70 000 homes per system. The government has received 128 cable franchise applications from various municipal cities.

The cable systems will be at least 12 channels and, depending on the erosion, may use coaxial cable. The three public networks (TFI, A2 and FR3) are must-carries for each system and 15% of each network's capacity must go to local programming under the control of the SEM. Foreign programming restrictions apply, limiting these channels to one-third of the total network capacity. Advertising rules are to be identical to those on the broadcasting networks and films must have been released for at least 2 years, to protect the cinema industry and support the local film production industry. Further, they cannot be transmitted at peak times. A share of the network receipts (30%) is to be invested in programming or in the purchase of new programmes.[13]

There are several major cable experiments planned in France. The mayor of Paris, Jacques Chirac, plans to connect 1 million homes with a combination of fibre optics and coaxial cable by 1992 at a rate of 150 000 households per year. His plan is to target districts to construct coaxial networks by 1986, moving later to fibre optics. He is also using private finance and private construction, rather than construction by the PTT.

In Biarritz and Montpelier, the French PTT has undertaken an experiment to gain experience in the installation, construction, and maintenance of fibre optic networks. The project began construction in 1980 and should provide 1500 homes with 15 television channels, stereo channels, and interactive services.

In Lille, a semi-public network is being installed and operated by the PTT, TDF, and the urban community. The experiment is to determine price levels for programmes and services in addition to testing the technical capabilities of the fibre optic network.

In total 8 cable systems had joined the experiment by late 1985 and a further 12 proposed to sign by the end of 1985. Already coaxial cable is being used in place of fibre optic to economize. By 1995, France plans to have cabled over 50% of the country. Although it is difficult to know which area of new media France is concentrating upon—DBS, new terrestrial television networks or cable—the 1985 telecommunications budget for cable in 1986 was an increase of 50% over the previous year's commitment.

In a major change in French policy, President Mitterand took the first steps in January 1985 to legally permit private programmers in France. The proposal is to allow up to 80 local stations connected to one or two private national networks run by a consortium of programming companies.[14] Programmes would be relayed from a central broadcasting station and then retransmitted locally. Each major urban area would be allocated one service under the existing RF (radio frequency) allocation. Large conglomerates, such as Television Monte Carlo/Europe 1, Hachette, RTL and the Hersant Group, applied and the first network was awarded to a consortium of Italian and French interests (Chapter 2). This policy initiative throws into turmoil the orderly development of cable and DBS in France. One option is to grant private television licences to municipalities which are already cabled (or committed to cable) so that some synergy is maintained in the broadcast policy.

6.3.6 WEST GERMANY

In West Germany, only 5% of the viewers were hooked to cable networks and 9% to MATV systems when a policy was announced in 1982 for the Deutsche Bundespost (DBP) to install wideband cable. With this announcement came a budget increase from 400 million to 1 billion DM for the DBP. This was to continue throughout the eighties to enable 1 million households per annum to be cabled. In 1984, cutbacks and rationalizations were introduced which effectively halved the growth rate of, but not the commitment to, cable.

The DBP programme was based upon a recommendation of the Federal Commission for the Development of the Technical Communications System in 1974, which proposed several pilot cable schemes in 1975. In 1978, the Länder and federal authorities agreed that networks should be installed in Dortmund, Ludwigshafen, Munich, and Berlin to be used for cable television trials and managed by the DBP. The projects are financed out of a contribution from each of the ARD regional stations and a levy, the Kabelgroschen, which was added to all television licences in Germany to fund the four projects. Private companies are allowed to participate in the cable operations. In Berlin, Dortmund, and Munich, the Länder broadcast organizations are to provide the programmes. In 1982, installation of the trial networks of optical fibres in Berlin, Düsseldorf, Hamburg, Hanover, Nuremberg, and Stuttgart began.

140

Systems in Munich and Ludwigshafen began in 1984, Dortmund's project was launched in mid 1985, and the Berlin network began in August 1985. Dortmund originally stated that it would not carry private television signals, being a public network, but that policy has been reviewed opening the door for SAT-1, Sky Channel, Music Box, and the like. This follows a rethink by the SPD Government in late 1984 and a decision to 'allow' private media.

The Bigfon experiment is the most sophisticated, incorporating two projects in seven cities. This network is constructed of optical fibre and is designed to test text and data services, improve television transmission, picture phones, and stereo.

Further progress has been made at Ludwigshafen which had 20 000 connections by September 1985 and 10 000 households waiting to be connected. It is a coaxial system designed to carry 22 television channels and stereo channels. Pay-television is under experimentation and is seen as the primary revenue source. Ludwigshafen provides an outlet for the pan-European satellite-delivered cable services. A related project at Mannheim will be connected by a fibre optic link and is to pass 10 000 homes.

The cable system in Munich is planned to pass 50 000 households. Pay-television is under experimentation and the Munich system plans eventually an offering of up to 60 television programmes.

The West Berlin cable network is based on existing coaxial cable, providing text and satellite-distributed services. The network is expected to pass 95 000 homes by the end of 1986.[15]

In 1984, the Bundespost experienced a slow take-up rate on its cable systems. Consequently, it began to 'subsidize' this programme to encourage cable take-up. In existing multiple dwellings, connection charges were reduced from 400 DM per connection to 3000 DM per 10 connections in a single building, and 20 DM per additional connection up to a maximum of 5000 DM. For new dwellings, the rates are even lower, being 2000 DM for 10 connections, 15 DM per additional connection, and a maximum of 3560 DM.[16]

6.3.7   UNITED KINGDOM
In 1982, a report was issued by the Hunt Committee, which had been established in the United Kingdom to study cable policy for the United Kingdom.[17] The basic findings allowed great flexibility with cable development, advising that there should be minimal regulation and control, including no limit to the number of channels offered, unrestricted finance from advertising or sponsorship, no obligations regarding content or balance and mix of programmes. Organizations currently operating a cable franchise would be forced to replace it within 5 years with a multi-channel

network. A new Cable Authority was established to oversee the new industry, and it began its jurisdictional duties in December 1984.

The government granted licences to 7 companies to provide pay cable services in 13 cities as an experiment:[18]

Rediffusion	Reading
	Pontypridd
	Hull
	Tunbridge Wells
	Burnley
Radio Rentals	Swindon
	Medway Towns
British Telecom	Milton Keynes
Philips Cablevision	Tredegar
	Northampton
Visionhire Cable	London
CableVision	Wellingborough
Greenwich Cablevision	Greenwich

The licences were granted for 2 years in an experiment to determine the feasibility of pay services. The results of the experiment were favourable, so the government proceeded with plans to extend cable by granting a set number of franchises.

In 1983, the Home Office took applications to establish new cable systems and to upgrade old systems. At that time, the United Kingdom had about 2.6 million cable subscribers or about a 14% level of penetration. Most systems are of four-channel capacity to accommodate the four government-operated broadcast services.

The franchises were to be granted to areas of about 100 000 people. There were 13 successful new applicants of the 137 that applied:[19]

	Homes passed
Aberdeen Cable Services (Aberdeen)	91 000
Clyde Cablevision Ltd (North Glasgow)	100 000
Ulster Cablevision (Belfast)	100 000
Merseyside Cablevision (South Liverpool)	100 000
Coventry Cable Ltd (Coventry)	100 000
Swindon Cable Ltd (Swindon)	53 000
Rediffusion Consumer Electronics (Guildford)	22 000
Westminster Cable Co. (City of Westminster)	73 000
Windsor Television Ltd (Windsor, Slough, Maidenhead)	84 000
Cable Tel (London Borough of Ealing)	100 000
Croydon Cable TV Ltd (London Borough of Croydon)	98 000
	921 000

Once the franchises were granted by the Home Office, it was necessary for British Telecom to grant a licence for guarantee of operation. There was considerable delay of a licence guarantee to the franchises, up to a full year after their applications were approved. Other delays also arose to slow cable development: non-exclusivity of local business services, practical difficulty with local authority planning, technical problems with cable installation, and other regulatory issues. By July 1985, Swindon had completed its first phase and was experiencing a very slow take-up rate; Aberdeen, Westminster, and Coventry had also commenced Croydon, Clyde by October 1985; and Windsor by December, 1985. Ulster, Merseyside, Ealing and Guildford was at a standstill although hoping to start by 1986.

While most of the applicants to upgrade old systems were approved, the costs, delays, and lack of government commitment in the United Kingdom have taken their toll. British Electric Traction (BET) has sold its cable interests (Rediffusion) to Robert Maxwell, Visionhire has withdrawn from new cable ventures and is closing down its old systems, the Plessey (UK) and Scientific-Atlanta (USA) consortium to supply equipment has been disbanded, as has the Racal (UK)—Oak (USA) consortium, and BICC is scaling down its cable activities. As a result, it is unlikely that the level of cable penetration will grow in the near future because any new growth will be offset by the loss of the old systems.

Despite these setbacks, and the fact that there were only 142 000 households connected to cable by July 1985 (most from the old Rediffusion networks), the newly established Cable Authority has plans to proceed with a systematic cabling of the United Kingdom. It called for tenders for five new franchises in February 1985 (Bolton, Cheltenham and Gloucester, Wandsworth, West Surrey and East Hampshire, and London Docklands/Tower Hamlets), a further three in June 1985 (Camden, Edinburgh, and Southampton), and proposes to repeat this process at frequent intervals.

6.3.8 SWEDEN

The government has acted on the recommendations of the Mass Media Commission (August 1984) and, in Sweden, it is possible for private companies to build and operate cable systems under a cable experiment project lasting through 1986.[20] There are 30 such experiments under way, although most of the investment in the cable itself comes from Televerket (the PTT), who see cable as an integral step in a broadband interactive network for Sweden.

The cable operator is responsible for programming on the system. Key restrictions are: (1) no local advertising permitted at all, consistent with the land-based broadcast policy, and (2) satellite channels beamed into Sweden can be relayed only if government permission is given and if the channel is already being relayed in its country of origin. This latter restriction allows foreign channels into Sweden which *do* carry advertising, such as Sky Channel.

The position of pay-television on the cable systems is not clear. SVT, the public broadcaster, proposed a pay-channel, but this was unacceptable to the government. Nordisk Betal-TV AB (NBTV) then proposed a 60 hours per week pay service initially through terrestrial transmission, then via Tele-X, the Nordic DBS service.

Esselte Video, in a deal with UIP, proposed to offer a pay service on PTT cable systems. The government has set up a committee to look into Sveriges Radio's future, and recommendations covering this, as well as advertising and pay-television, are expected in 1985.

6.3.9 AUSTRIA

Austria has plans to expand its base of cable systems which were originally installed to improve reception in the mountains. At present, local and regional companies are putting together private consortia to provide cable systems. The largest single system is Telekabel which is being built in Vienna. It will eventually reach 600 000 households, 450 000 to be cabled by the end of 1985.

There has been slower growth than anticipated in the development because public interest is low. The primary reasons are a high connection charge plus monthly subscriber fee. As viewers can already receive programmes from West Germany and Switzerland, the cable package is not particularly easy to market.

6.4 Cable programming

The increased regulatory flexibility and promise of profit based on the US experience has prompted numerous consortia to offer programs for distribution to cable systems. Unlike the USA, which has a large, homogeneous audience, European programme providers must consider the cultural and language diversity of its potential audience. These services are considered in detail in Chapter 8.

6.4.1 SERVICES—BY END 1985

Cable services take two main forms: those pan-European channels distributed via satellite and national channels distributed either via satellite, terrestrially, or directly from the head end. Pan-European services are comprised of three basic channel types: new 'super stations', rebroadcast channels, and specialized channels, targeted towards a pan-European audience. The national services are comprised of national channels, distributed terrestrially and via satellite and local channels, which are distributed from the cable head end, through the use of a computer or video cassettes. The market is evolving rapidly, there being 50% more channels in Europe in 1985 than before the launch of Eutelsat 1-F1.

6.4.2 PAN-EUROPEAN CHANNELS

The first grouping of programmes is comprised of pan-European channels distributed via satellite. These are primarily general entertainment television services that are viewable with a low comprehension level of the language. They are aimed at a European-wide audience.

The programme schedules of these channels consist of a combination of music, light entertainment, and drama series. The programming on these channels is similar to the superstations which broadcast in the United States. They are primarily advertiser supported, with no subscriber fee. Those channels currently under operation are located on Eutelsat 1-F1:

Channel	Language	Satellite	Start-up
Sky Channel	English	Eutelsat 1-F1	April 1982
SAT/1	German	Eutelsat 1-F1	January 1985
Europa TV/Olympus	multi-lingual	Eutelsat 1-F1	October 1985
RTL Plus	German	Eutelsat 1-F1	August 1985

These channels are broadcast via Eutelsat 1-F1 a satellite with a signal that can be received throughout Europe.

6.4.3 REBROADCAST CHANNELS

The second type of programmes are those which are oriented towards a pan-European audience but with content which is a rebroadcast of programming from terrestrial channels. Some of the channels, such as TV-5, broadcast programmes that require a low level of comprehension of the language, while RAI is targeted at a very specific language group and is a relay service for the Italian public broadcasting network. 3-SAT is advertiser supported and free of charge to subscribers, whereas TV-5 is subsidized by the various French language broadcasters in Belgium, Switzerland, and France. Like the other pan-European channels, they are transmitted via Eutelsat 1-F1. These rebroadcast channels include:

Channel	Language	Satellite	Start-up
TV-5	French	Eutelsat 1-F1	January 1984
3-SAT	German	Eutelsat 1-F1	December 1984
RAI	Italian	Eutelsat 1-F1	June 1984

6.4.4 SPECIALIZED CHANNELS

The third group of programme channels are those which specialize in content, but are still targeted towards a general European market. They are transmitted via pan-European signals and are advertiser supported. These specialized channels include:

Channel	Language	Satellite	Start-up	Type
Music Box	English	Eutelsat 1-F1	July 1984	Music
New World Channel	Norwegian	Eutelsat 1-F1	January 1985	Religious
World Public News	English	Eutelsat 1-F1	July 1985	News
World Net	English	Eutelsat 1-F1	April 1985	General US TV
CNN	English	Intelsat	September 1985	News

6.4.5 NATIONAL CHANNELS

Other satellite-distributed services are nationally oriented to a particular country. One group is comprised primarily of pay services which can be marketed more easily at a national level, and often include films and movie products, the rights for which have been sold historically for language and national markets in Europe. Most of these services are distributed via satellite. They include:

Channel	Language	Satellite	Start-up
Mirrorvision	English	Intelsat VA-F10	March 1984
Premiere	English	Intelsat VA-F10	September 1984
Teleclub (Paysat)	German	Eutelsat 1-F1	May 1984
FilmNet/ATN	Dutch/English	Eutelsat 1-F1	March 1985

A second group of services are terrestrial pay channels which are offered to subscribers who purchase or rent special antennas and decoders to pick up the signals. These services include:

Channel	Language	Start-up
Telecine Romandie	French	November 1985
Canal Plus	French	November 1984
Arts Channel	English	September 1985

A third national channel category comprises specialized channels distributed by satellite to a national audience; they are advertiser supported. These include:

Channel	Language	Satellite	Start-up	Type
Children's Channel	English	Intelsat VA-F10	March 1984	Children's
ScreenSport	English	Intelsat VA-F10	March 1984	Sport
Lifestyle	English	Intelsat VA-F10	October 1985	Women's, daytime programmes
STV	French	Telecom 1	November 1985	Music, films, entertainment

6.4.6 LOCAL CHANNELS

The final group of cable channels are not distributed by satellite at all, but rather through cassettes or computers situated at the cable head end. These are specialized channels, targeted towards a narrowly defined subscriber group in a particular language, and they do not need the distribution offered by a satellite. Because new programme regulations for cable are as yet the most flexible in the United Kingdom, many of the planned channels are English services. The German Länder have recently introduced legislation that will allow local services to develop on their cable systems. As a result, many groups are in the process of developing services to fulfil local needs.

6.4.7 DEVELOPMENT

The emergence of these programmes is similar to the development experienced in the United States. Shake-outs have already occurred, as the subscriber base is too narrow to support all the proposed channels. It is likely that further mergers between channels will take place, and other channels will go out of business entirely. The evolution of cable programming in Europe will depend heavily upon the speed at which Europe is cabled in the next several years.

6.5 Hardware provision

Many of the present cable systems need to be upgraded to provide more channel capacity before they can become profitable outlets for programme providers (Table 6.3). The cost to upgrade an average system in Europe is estimated to be about $150 per household.[21]

In the United Kingdom, some attempts were made by cable providers to upgrade their systems through the installation of aerials to receive the terrestrial signals, freeing the cable system to supply other entertainment services. Because the subscription uptake on the new systems has been so poor, many of these upgraded systems have closed down. The combination of limited channel capacity on old systems and unprofitable upgraded systems indicates that the old systems will not be a good source of subscribers. As a result, the focus of cable development is on the newly constructed systems.

6.5.1 CABLE CONFIGURATIONS

The main hardware configurations that exist for cable are tree and branch and switched star. In older European systems, a loop system is used, where one cable was strung from one house to the next in series. It is difficult to add new subscribers to such a system because the loop cannot be easily

Table 6.3 Average channel capacity on cable systems in Europe, 1985*

Country	Average channel capacity	Comments
Austria	12	The Vienna system is an exception: it has 18–20 channels
Belgium	20	
Denmark	4–6	Broadband wiring is under way to connect the cable and, mostly, MATV systems
Finland	8	HTV in Helsinki is an exception: it has 6 at present but is being upgraded to 20 channels
France	12–20	Pilot projects are 20 channels
Ireland		
Netherlands	12	Large cable operators are upgrading to 20 channels
Norway	6	
Sweden	12	
Switzerland	12	
United Kingdom	4	Rediffusion Network
	20	New systems
West Germany	12–20	New systems

*Taken from interviews with cable operators across Europe.

disconnected. For this reason, the system is not widely used and the Netherlands has even banned this network design since 1978.

The tree-and-branch system is the most widely used configuration because it is the least expensive to install. Standard coaxial cable is used and it has little electronic switching, which limits interactive possibilities but lowers relative costs. One major trunk cable extends from the cable head end, and repeatedly branches off like a tree. The full range of programmes is available to all subscribers, but the signal gets weaker near the end of the branch and the capacity of the system is limited to the most narrow branch. The major disadvantage of this system is that it needs bulky and expensive equipment at the subscriber's home, which increases the cost to the individual, and requires frequent amplifiers to boost the signal.

Star and switched-star systems cost between 10 and 20% more to install than tree-and-branch systems.[15] The increase in cost is due to the installation of more local points of distribution, each feeding a fewer number of subscribers. Only the trunk or main line needs to carry all of the services and an electronic switching system is installed at each of a number of nodes or distribution points. These nodes act as programme selectors, and the final link to the household only carries three or four channels. The local distribution points remove some of the pressure from individual lines and make it cheaper to cable each home, as it eliminates the need for expensive home equipment. The total star system, however, is not cheaper than the tree and branch.

The star systems are more flexible than tree-and-branch systems, and more interactive services are possible. However, tree-and-branch equipment is widely available and there is considerable experience in the industry with the use of this equipment.

Most of the new systems will be installed with coaxial cable which is bulky and heavy, and is very susceptible to outside interference.

The major alternative to coaxial cable is fibre optic cable, which costs about $290 to $300 per kilometre to manufacture,[15] making it considerably more expensive than coaxial cable. It is made with a core of glass, plastic, or ceramic material which transmits signals in the form of light pulses. It has great flexibility, a large carrying capacity, and interactive capabilities, and it does not receive interference like electrical transmissions.

As fibre optics are so much more expensive than coaxial cable, and their immediate advantages may not be apparent to consumers, most of the planned systems will be tree and branch, using coaxial cable. However, government-sponsored experiments to examine the most sophisticated cable services are using switched-star configurations, with fibre optic cable.

6.5.2   COST
There are several possible hardware types for cable systems which vary in price and performance. In areas where cable must be installed underground, primarily in populous areas, costs are the highest, between $500 and $750 per home passed. An average system costs between $1 and $3.5 million to construct, and construction and hardware usually comprise 80 to 90% of the costs.

The Luton and Dunstable model (EIU) is indicative of costing for the densely populated areas of Europe:[15]

Cable purchase and installation:	$23 000 per mile
Running costs	$575 per mile
Connection per subscriber:	$175
Decoder	$144
Head end and operational facilities:	$1.3 million in the first year

In the United Kingdom, where franchises have begun to operate, a combination of ways have been proposed to raise capital. These include: lease of cable equipment by UK resident parties (capital allowances for franchises granted after the first round will increase lease rentals by 10% for a 12-year franchise lease), capital allowances for cable ducts and installation costs (however, first-year allowances on industrial buildings have been eliminated), and a combination of investors and deep discount securities. Deloitte Haskins and Sells has illustrated a typical cash flow situation for a UK franchise where 96 000 homes are passed, a maximum penetration of 50% of the basic services is achieved by the seventh year of operation, subscribers pay $140 yearly for basic services, and some advertising income is earned (Table 6.4).

Table 6.4 Internal rate of return[22]

Franchises beginning	With tax relief (%)*	Without tax relief (%)†
Pre-Budget	13.51	7.85
April 1983	9.15	7.06
April 1984	8.03	6.68
April 1985	7.53	6.50
April 1986	7.44	6.47

*Tax losses used by group relief surrendered; payment for these losses representing tax saved by the investors will be made to the cable company.
†Tax losses cannot be surrendered and are carried forward in the operating company.

It is apparent that, as time progresses, the franchises have less opportunity to become profitable. The changes in capital allowances together with the abolition of tax relief on limited partnerships (in the Finance Bill, 1985), which limited relief to the amount of the initial investment, are making cable an unattractive proposition. The low level of activity, as seen in the United Kingdom by 1985, has stimulated an industrial lobby to improve the tax relief.

6.6 Cable growth

The early expectations of rapid cabling of Europe received a severe jolt in the light of the economic reality of the eighties. In the United Kingdom, for example, the 1984 retrofit programme by Rediffusion, a programme to encourage 1 000 000 households on old twisted-pair cable systems to subscribe to a package of four satellite-delivered channels, Sky Channel, Music Box, The Movie Channel (a pay channel which went into liquidation in May 1985 and is being revamped as Mirrorvision, a channel owned by Robert Maxwell, the UK publisher), and Screensport, was a disaster. Not only was the take-up rate slower than predicted, but the feasibility of the exercise had been grossly underestimated. The result: 140 000 households (at most) by July 1985 and a prediction of only 200 000 by the end of 1985. Further, not one of the eleven interim cable franchises in the United Kingdom began operation on schedule. Financial difficulties and uncertainties dominated the industry.

In West Germany and France, while both governments reaffirmed their commitments to cable, expenditure levels and cabling rates were both revised downwards significantly in 1984. The Bundespost is now considering a mixed media model: cable for urban and DBS for rural West Germany. The local media laws have taken much longer to implement and of the four pilot cable projects, Berlin did not start until late 1985. With uncertainties about its DBS programme, France has become more buoyant (again) on cable.

The result in Europe is one of an uncertain eighties for programme providers looking to cable system outlets for their services (Table 6.5).

Table 6.5   European cable antenna television market, 1985–90*

Country	Cable households as a percentage of TV households		Five-year growth (%)	1990 cable as percentage of TV households
	1985	1990		
Austria	0.60	0.9	50	33
Belgium†	2.80	2.9	4	88
Denmark	0.32	0.6	88	27
Finland	0.18	0.4	120	24
France	0.75	2.0	167	10
Greece	—	—	—	—
Ireland	0.24	0.3	25	33
Italy	—	—	—	—
Netherlands	2.58	3.4	32	71
Norway	0.26	0.4	54	29
Portugal	—	—	—	—
Spain	—	—	—	—
Sweden	0.20	0.5	150	15
Switzerland	0.95	1.3	37	57
United Kingdom‡	1.95	3.5	79	17
West Germany	1.47†	5.0	240	20
Total Europe	12.30	21.9		22

*Forecasts are derived from published studies and conversations with cable operators and PTTs in Europe.
†Includes Luxembourg.
‡Includes all of the old four-channel systems used to carry television in hilly terrain. There are only 140 000 households receiving the 'new' cable channels.

Agreement on where cable (or SMATV) is going is not forthcoming. For example, the following are published 'forecasts' of European cable penetration in 1990 made between 1981 and 1984, giving the sources and number of households (in millions):

Mackintosh[10]	36
CIT (1982)[21]	16–27
CIT (1984)[23]	14–20
Link[24]	22
Ogilvy and Mather[9]	24

The most recent figures from CIT Research indicate that forecasts for investment are increasing while subscriber forecasts are decreasing. They predict that the overall growth of cable will proceed at about 10% annually, from a much-reduced subscriber base.[23]

Basic services exclude premium service uptake by several per cent. An estimated $880 million will be spent to provide programming to subscribers while $1.1 to $2 billion will be spent annually on cable construction and installation by 1994.[23] Subscriber revenue is expected to reach $2.2 to $4 billion. Optimistic forecasts are that advertising revenue will be $4 to $5.5 million by 1994. It is more reasonable to assume that cable will be a secondary medium to the networks for local advertisers. But super channels such as Sky Channel will succeed in generating considerable revenue. Even if all of the predicted potential demand for advertising went to cable (Chapter 3), our view is that the total market is likely to be at most $1 to $1.5 billion (at mid 1985 exchange rates) or between 25 and 40% of the current European television advertising expenditure. It is apparent that cable is off to a much slower start in Europe than was predicted several years ago.[23]

6.7 Satellite master antenna television (SMATV)

The orderly evolution in Europe of cable in urban areas followed by DBS in rural areas has taken a temporary setback. All DBS projects have been subject to some delay (Chapter 5) and most new national cable programmes have been scaled down, primarily because of the economics of the proposed ventures. It is becoming clear that Europe will not follow the US path and achieve 60% of households passed by cable in the short term. Instead, the role of the master antenna system, or 'cable island', will play a much greater role in the European new media scene.

Thus, SMATV (satellite master antenna television) is emerging as the wildcard—the possibility of direct reception from FSS satellites to communal dwellings (flats, apartments, businesses) and, although not strictly SMATV, hotels, multiple-dwelling units, hospitals, and clubs. The importance of SMATV has grown with the realization that cable will be slower, more costly, and probably have lower take-up rates than originally believed.

In fact, SMATV may be a way to stimulate the cable market and help to provide a viable market for programmers.

The size of the SMATV market is unknown, although some estimates have been made. Our forecasts are of a market potential estimated to be between 25 and 28 million households connected to MATV systems of varying channel capacity and technological sophistication (Table 6.6).

Table 6.6 MATV systems and households*

Country	MATV systems, 1983 (thousand)†	MATV households, total market potential (million)	Estimated satellite MATV connected households by 1990 (million)
Austria	80	0.6	0.25
Belgium	5	0.1	0.04
Denmark	10	0.6	0.12
Finland	15	0.4	0.04
France	450	6.3	2.20
Greece	—	—	—
Ireland	—	—	—
Italy	700	5.8	1.10
Netherlands	14	0.8	0.28
Norway	5	0.25	0.06
Portugal	—	—	—
Spain	20	2.5	0.50
Sweden	50	1.0	0.15
Switzerland	60	0.65	0.25
United Kingdom	10	1.5	0.53
West Germany	400	7.5	2.63
Total	1819	28.0	8.15

* The figures exclude the hotel and hospital markets.
† From Mackintosh International.[10] There are no reliable sources and these numbers include MATV systems of all sizes.

There are two barriers to rapid growth of the SMATV market. The first is practical economics. Many of the systems are for groups of less than 10 dwelling units. They use old twisted-pair copper cables and are wired to receive only the national broadcast channels (the must-carry channels). Further, they have very limited capacity, usually four to six channels at most. To receive satellite signals would require not only the purchase of an antenna, receiver, and decoder (or decoders, as there is not a common standard as yet), but possibly a complete rewiring of the dwelling unit.

The capital costs for an SMATV system distributing three channels from Intelsat could be as high as $13 500: the antenna, LNA, and demodulators cost $5500, the modulators and amplifiers $1700, the installation $100, and the encoders $5300.[25] These costs are expected to fall rapidly as Europe's hardware manufacturers enter the market for providing hardware to hotels, apartment blocks, and multiple-dwelling units. Already the prices that are being quoted have fallen. For example, in the United Kingdom where the SMATV marketing first began in Europe, one-channel systems are available for $2000 (the dish, the low noise converter, and the receiver) and four-channel systems for around $4000.[26] Thorn has bundled its programmes together—Premiere, Children's Channel—with ScreenSport to offer a Galaxy package. A second dish is needed to pick up Music Box and Sky Channel. A similar strategy is being developed in Germany where RTL-Plus, SAT-1, Teleclub, Music Box, and Sky Channel are all on the Eutelsat 1-F1 satellite.

There is also the pragmatic problem for European urban dwellers of: (1) getting permission from local planning authorities and (2) finding a suitable location to install a 1.8 to 3-metre antenna. Our estimates are that these constraints alone would reduce the size of the potential market by 50 to 60%. Of course, once mass market DBS is available, these dwelling units may be candidates for that service.

The second constraint is even more fundamental. Satellite-delivered programming to cable, MATV, or households may require a licence. In Switzerland and Finland, SMATV is legal, although it requires a licence by the PTT. In Austria, government approval is needed while in Spain, Ireland, and Denmark, SMATV reception is not legal.

In Sweden, it is legally possible for SMATV services to operate for private purposes. Licences are required for the reception of satellite signals, but systems with less than 50 connections are exempt from the licensing requirements. In France, it is theoretically possible to obtain a PTT licence to receive satellite-delivered channels, but the PTT is not yet issuing licences. Proposed regulations would relax licensing restrictions for satellite reception and put the authorization of SMATV licences under the authority of the Haute Autorité Audio-Visuelle. As an interim measure while cable development progresses, restrictions on MATV are being lifted in order to stimulate subscribers for the cable services.

In the United Kingdom, the government introduced Europe's first major piece of legislation for SMATV in May 1985. There are three cases: individual premises, existing systems, and new SMATV systems. The first case is easiest. Individual premises, and this includes hotels, holiday camps, apartment blocks, and even hospitals, require a licence under the Wireless Telegraphy Act 1949 in the same way as an individual wishing to receive a television programme.

In the other two cases it will be necessary to get licences under the Cable and Broadcasting Act 1984 and the Telecommunications Act 1984. For systems which are licensed at present, such as the relay networks, an automatic 3-year extension of the existing licence will be granted provided that the region has not since been advertised for a cable franchise. The must-carry rules (which includes the two BBC channels and the two ITV channels) will still apply.

New franchises will be granted normally for a 5-year period. Again the three licences (Wireless and Telegraphy Act from the Department of Trade and Industry to receive the low-power satellite signals, Cable and Broadcasting Act to diffuse the programme service, and Telecommunications Act to run the system) will be required. For new systems, the must-carry rules are extended to include existing broadcasters as well as any DBS channels originating in the United Kingdom.[27] In November 1985, the Cable Authority granted the first four licences to SMAN operators to provide cable channels for up to 17 000 households.[28]

In West Germany, the DBP has liberalized the reception of signals from satellites for MATV systems and individual households. While the DBP continues to influence the installation and control of antennas for cable systems, the July 1985 decision basically deregulates the reception of signals elsewhere. A one-time only licence fee is payable, but will be granted upon submission to households not in broadband cable networks, and the fee has been set to stimulate penetration: 25 DM for individual households and a similar fee for MATV households up to a maximum of 25. Thereafter, the price drops to DM15 per household.[29] This decision does not affect media laws, which are formulated by each Land, but the decision makes it difficult to prevent reception of satellite services throughout Germany.

It is hoped by those involved with cable provision of services and operators of franchises that the interim MATV systems will develop an audience for the services so that they will survive, while preparing them for the introduction of cable franchises in their areas. However, it is also possible that the introduction of MATV systems may eliminate some of the urgency felt by the programme providers to construct more systems, which would allow the terrestrial systems to gain a stronger foothold.

The total number of households connected to MATV systems is unlikely to increase. The barriers to growth of systems capable of receiving satellite-delivered programming are:

1. *Technological*
   (a) The size of the antenna to produce acceptable picture quality is too large.
   (b) The existing systems have insufficient capacity to carry additional channels.
2. *Economic*
   (a) Upgrading the system is too costly.

(b) The satellite receiver equipment is too expensive.
3. *Regulatory*
   (a) Planning permission by local authorities is not forthcoming.
   (b) Licences to receive satellite signals from fixed satellite services are not issued.
4. *Utility*
   The array of programming (in any given language) on any one satellite is not of sufficient interest to warrant the expense of installation.

Applying the above factors to the estimated total MATV market, excluding hotels and hospitals, we get a forecast of 8.1 million households likely to take MATV by 1990. (This forecast assumes that there are no major national political barriers to downlinking; see Table 6.6.) The size of this market is directly related to the rate of cabling and the early success of DBS. However, it offers Europe the best chance of a support market to the existing cable systems.

6.8   Uncertainties in development

While it is apparent that cable is moving forward in Europe, there are many uncertainties which could further hinder its development. Cable has remained primarily under government control. While most countries are in the process of changing their regulations to accept some form of media, the establishment of new legislation has been slow. Most governments are afraid to invest large amounts of capital and time if they are not entirely certain that cable will be profitable. Europe is densely populated and cable installation underground is extremely expensive. As cable construction appears to be a rather risky investment, any private franchises will need to be funded through a combination of investors. Capital allowances in countries such as the United Kingdom are poor, adding further risk to cable as an investment.

There are also other competitive technologies, such as DBS and medium-powered satellites, that may prove to be a more cost-effective means of distribution. Government funds are not readily available for all of these projects, so they are slow to determine the best means of allocation for their funds. There is a belief by some new media analysts that satellite services will increase their technological capabilities and become more cost effective, making cable an out-of-date, expensive distribution method. Some European countries are committed to DBS as well as their full efforts behind cable.

A serious consideration for cable is the actual demand for the services. VCR penetration is very high in Europe, and software distribution is easily accessible. As a primary reason given for consumers to take cable services is demand for movies, the VCR industry may adequately satisfy this demand, and consumer interest in cable could become increasingly low as VCR penetration increases. HBO and Showtime, pay-movie channels, have experienced considerable churn in the United States in 1985 as VCR penetration has increased. Take-up rates for the first cable systems have been lower than anticipated, which threatens an already low subscriber base for new services.

Most governments have focused upon the interactive services as the primary reason for cabling. However, the US experience has demonstrated that consumers are not particularly interested in the services offered by interactive capabilities, and the cost to install and maintain these networks makes the services unprofitable. Considerations over network type have also caused delays in the actual implementation of cable.

6.9   Conclusion

Cable has been a topic for extensive discussion in Europe over the past several years. Many governments have implemented legislation to support a cable industry. The cable programme services range from general entertainment satellite delivered channels with few language skills necessary, targeted at the entirety of Europe, to highly specialized services such as computer games, distributed from the cable head ends. Because the subscriber base is still small, it is unlikely that all these services will be profitable, and many may merge or disappear altogether. It will not be profitable to upgrade many of the old systems, so they will not provide good outlets for the new programme services, which will need to rely on the development of the new cable systems.

Except in the Netherlands, Belgium, and Switzerland, cable penetrations are relatively low. The large capital-intensive investment required up-front and the emerging competitive media (VCRs, DBS, and low-power television) are challenging the economic viability of cable. With the exception of the Bundespost (in West Germany), no country has given unequivocal support to cable either through the public sector infrastructure or the marketplace. The result is uncertainty on the ground and frustration for the programme providers.

An interim solution may be through SMATV. Over 20% of European households are connected to MATV and with the decline in cost of the electronics for satellite antennas, this route may prove to be economical. The satellites to deliver programming are already in orbit.

The next 3 to 5 years will be significant ones for European cable as the 'big three'—the United Kingdom, France, and West Germany—begin cabling in earnest.

References

1. Derived from *CableVision* and A.C. Nielsen data in the Economist Intelligence Unit, *Cable Television in Western Europe*, EIU Ltd, London, 1983.
2. Derived from *Broadcasting/Cablecasting Yearbook 1984*, Broadcasting Publications Inc., Washington, DC, 1984.
3. Alastair Tempest, 'The Witches' Brew', *Journal of Advertising*, Volume 1, 1982, pp. 143-55.
4. *CableFile/82*, Titsch Publishing, Denver, Colorado, 1982.
5. 'Wiring America', *CableVision*, 10 January 1983, pp. 25-53.
6. *Television and Cable Factbook*, Cable and Services volume, Television Digest Inc., New York, 1983.
7. 'CAB Cume analysis', Cable Advertising Bureau, New York, November 1983.
8. 'Cable industry growth chart', *CableVision*, 28 May 1984, p. 136.
9. Ogilvy and Mather Europe, *The New Media Review 1984*, Ogilvy and Mather Europe, London, 1984.
10. Mackintosh International, *Satellite Broadcasting*, Volume 3, Mackintosh Consultants Ltd, London, 1981.
11. Nick Snow, 'Film channels push forward', *Cable and Satellite Europe*, December 1984, pp. 21-2.
12. 'Le câble contre vents et marées', *Le Monde*, 5 September 1985, p. 16.
13. Jack Monet, 'French cable limits imports via 30% quota', *Variety*, 2 May 1984, pp. 1 and 175.
14. 'French "oui" for private TV', *Cable and Satellite Europe*, February 1985, p. 8.
15. *Economist* Intelligence Unit, *Cable Television in Western Europe*, EIU Ltd, London, 1983.
16. Simon Baker, 'Cabling the country in fifteen years', *Cable and Satellite Europe*, November 1984, pp. 33-5.
17. Commission of the European Communities, *Interim Report: Realities and Tendencies in European Television: Perspectives and Options*, Commission of the European Communities, Brussels, 1983, pp. 129-31.
18. Financial Times Business Information Consultancy, *Key Issues: Cable Television in the USA and UK*, FT Business Information Ltd, London, 1982.
19. 'The eleven pilot franchises in the UK-update', *Cable and Satellite Europe*, March 1984, p. 30.
20. 'Update on cable in Sweden', *Ogilvy and Mather Euromedia*, January 1985, p. 9.
21. Communications and Information Technology (CIT) Research Ltd, *Cable TV in Western Europe*, CIT Ltd, London, 1982.
22. Deloitte Haskins and Sells, *Cable Television: The Implications of the Budget Statement for the Development of UK Cable Television*, Deloitte Haskins and Sells, London, 1984.
23. 'CIT Survey II—governments must free cable', London, *Cable and Satellite Europe*, April 1984, pp. 14-17.
24. Link Resources Corporation, *Cable and Satellite Developments in Western Europe (1982-1987)*, Link Resources Corporation, London, 1982.
25. 'SMATV: the emerging market', *Media Monitor*, 14 March 1985.
26. 'Supplying the UK market', *Cable and Satellite Europe*, June 1985, p. 15.
27. 'Licensing of cable diffusion services', *Cable Authority*, London, May 1985.
28. A. Burkitt, 'First small-scale licences granted', *Broadcast*, 15 November 1985, p. 13
29. 'SMATV moves in West Germany', *Cable and Satellite Europe*, August 1985, p. 22.

7

Video cassette recorders, video disc players, and compact discs

7.1 Background

Until 1979, European sales of video cassette recorders (VCRs) totalled only 340 000 units, with the largest concentrations being in the United Kingdom, France, and the Netherlands. The year 1979 was the turning point; annual growth rates of over 100% were experienced for the following 3 years in many countries. The United Kingdom and West Germany led the way, and the Netherlands, Sweden, Norway, and Ireland had all passed the 20% household penetration level by the end of 1984.

Table 7.1 VCR growth in Europe. Forecasts of households

	1981 (thousand) *	1984 (thousand) †	1990 (thousand) †	1981 (%)*	1984 (%)†	1990 (%)†
Austria	60	140	642	2.2	5.0	23
Belgium	150	200	979	4.4	5.9	29
Denmark	80	220	572	3.8	10.4	27
Finland	15	80	371	0.9	4.6	21
France	440	1160	6 288	2.3	6.0	32
Ireland	—	90	296	—	9.8	32
Italy	55	370	1 552	0.3	2.0	9
Netherlands	—	380	1 677	—	7.7	34
Norway	90	220	521	6.4	15.7	37
Spain	110	1100	1 625	1.3	12.7	19
Sweden	285	500	1 319	8.4	14.7	39
Switzerland	—	160	500	—	6.8	21
United Kingdom	1550	5740	9 996	7.6	28.1	49
West Germany	1745	4500	10 937	7.0	18.1	44

* From *Broadcast*.[1]
† From Ogilvy and Mather Europe.[2]

VCR sales and penetration levels are expected to continue their growth, but not at the same rapid rate experienced during this 3- to 4-year boom period (Table 7.1). AGB Consumer Research and Euromonitor have both published surveys which indicate that sales growth is slowing in countries where VCRs have already reached high levels of penetration. This is particularly true in the United Kingdom and Germany, for example, where sales levels in the last half of 1984 caused alarm among distributors. Sales in the first half of 1984 in the United Kingdom were less than sales in the comparable period of 1983. Retailers have blamed some of the decline on the software market, suggesting that the significant lack of new titles is responsible. Rental firms, which account for a substantial number of VCRs in the United Kingdom, blame a reduced standard of living or at least a 'poorer market' from which to select now that significant penetration of the higher income groups of consumers has been attained. Based upon these recent trends, the upper limit to VCR penetration may well be nearer to 50% of television households, not the original predictions of 80 to 100%. Other European countries may experience a similar trend once a 25 to 30% penetration level is reached.

On the hardware side, Europe has been unique because three, rather than two, VCR formats have been competing for consumer acceptance – the two 'world standards', Sony's Beta and JVC's VHS format, and a European system, Philips-Grundig V-2000.

Philips-Grundig, a Dutch–German consortium, introduced their first domestic VCR, N-1500, in 1974, about the time the VCR market was beginning to open up in the United States. Consumers in the United Kingdom, West Germany, and the Netherlands, major users of Philips' products, began to purchase VCRs during 1976 and 1977. In 1977, Philips-Grundig made improvements on this model, and launched a new Long Play System, the N-1700, which was nearly obsolete before it was even introduced.

When the Japanese manufacturers began to distribute their VCRs in Europe in 1978 and 1979, Philips was in a relatively weak position. Their second model was too expensive to compete with the Beta and VHS formats from Japan, so Philips introduced a third model, the V-2000, which has remained their major model for Europe; the company has never been able to catch up with either Japanese format.

7.2   VCR—consumer use

VCRs are used by consumers in Europe for a variety of purposes. Time-shifting is the most popular: 70% record programmes while watching another channel; 69% record programmes at inconvenient times; and 40% purchase a VCR in order to buy or hire programmes that they have recorded.

VCR viewers tend to use the television more often than other television viewers, and they are more likely to subscribe to cable television and have an interest in other new media.

Many of the first VCR owners were hobbyists. Their profile was: higher-than-average income, fewer children than average, stable city dwellers, and aged over 50 years. The majority of more recent VCR owners, however, is quite different. They tend to own other consumer electronics equipment, are married with young children, and are employed full-time.

The local television programmes available and the level of colour television penetration significantly impact upon the VCR uptake. Where numerous films are broadcast on terrestial television, like Italy, rental of film videos is much lower than in other countries. Where a wide choice of television programmes are available, VCRs are used to time-shift and record while other programmes are on. Where local programming is limited in variety, and reception is not always good, as in Sweden, Finland, and Norway, there is a greater demand for pre-recorded programmes.

The use of VCRs to download software has been discussed by software suppliers in Europe but the only commercial venture so far has been in the United States: the ABC/Sony Telefirst experiment in Chicago. In this service, 3000 to 4000 subscribers received movies, exercise programmes, and children's short subjects over the broadcast airwaves between 01.00 and 06.00 in the morning. Three and a half hours of scrambled programmes were broadcast for six nights a week, at a cost of $17.95 to $25.95 per month, depending upon the number of programmes requested.[3]

The service was terminated in August 1984 with a $15 million loss. ABC concluded that 30 to 35% VCR penetration was necessary for downloading to be profitable. However, high VCR penetrations also facilitates a viable pre-recorded tape market. Even in the United States with less than 20% penetration of VCRs in early 1983, rental prices were as little as $1.00, and Telefirst only supplied four films per month. For price and convenience, Telefirst was not a viable option to the rental markets.

The United Kingdom and West Germany, have reached a level of penetration to contemplate downloading. However, video rental outlets are firmly entrenched in the markets with high VCR penetration in Europe, which makes entry into the market difficult for a downloaded service at this time. Decoders are necessary for use with a downloaded service; they are costly and inconvenient if a programme is not watched from start to finish.

A downloaded service necessitates the purchase of blank tapes. This market has been relatively slow in Europe and does not appear likely to change. The blank tape market also reduces proportionately as VCR penetration and rental markets expand.

Downloading has one major advantage in that it can utilize airwaves during low-cost transmission times. Programmers also feel downloading could be successful if first-seen television rights were procured for films for use in a pay-per-view window opportunity, a contentious and unproven medium in the US cable environment. Despite all of the commercial misgivings, programmers such as Thorn EMI believe a future for in-home downloaded services exists. However, it may arrive through specialized and business services, and by-pass entertainment altogether.

Finally, the use of VCRs could have a serious effect on the effectiveness of advertising on commercial television stations. 'Zapping' occurs when VCR users pre-record programmes and fast forward during the commercials during playback. The frequency of 'zapping' is thought to be quite extensive and increasing. Advertisers have become concerned because viewing figures upon which advertising rates are based may be radically different during the commercials and the programmes. Advertisers are now addressing this issue as a major concern because VCR growth and 'zapping' are expected to continue.

7.3   VCR growth

There are a number of factors which have influenced the growth rate of VCR penetration in each European country. In general, the wealthiest countries with more expendable income, such as Switzerland, Sweden, and Norway, have a high level of penetration, while poorer countries, such as Portugal or Italy, have the lowest level of penetration. Ireland, the poorest country in the EEC, is an exception, having a relatively high level of VCR penetration (Table 7.1).

Viewer choice of traditional broadcast television also affects the uptake level. The lack of choice of channels, or programmes of general interest, like films, has stimulated growth in Sweden and Switzerland. Industrial countries where VCRs are manufactured, such as the United Kingdom, West Germany, the Netherlands, and Sweden, also tend to have higher penetration levels.

The large rental market, which was easily adapted to include VCRs, appears to have had a significant influence in the achievement of high penetration levels quickly in both the United Kingdom and Ireland. Through low monthly rental fees, VCRs have been accessible to a wider number of people who might not otherwise be able to afford the capital outlay required for a VCR. In addition, UK television is the most 'entertainment-oriented' in Europe, so time-shifting is popular in both the United Kingdom and Ireland (which can receive UK programmes).

Other factors affecting VCR growth include marketing, advertising, and price reductions. Despite the restrictions on trade in electronics hardware in Europe, technological advances and economies of scale in production of VCRs have meant that price levels have continually decreased. VCRs have become more affordable.

The percentage levels of the three formats—Betamax, VHS, Philips—have fluctuated as well, but overall VHS has emerged as the market-determined *de facto* standard. VHS is the predominant format across Europe generally with 66%, while Beta is only 23% (Table 7.2).

In the German- and Dutch-speaking countries where the thrust of the Philips–Grundig consortia have their manufacturing units, the V-2000 has a higher level of penetration than in the other European countries; Austria (60%), West Germany (54%), the Netherlands (20%), and Switzerland

Table 7.2 VCR formats in use by country by percentage*

	VHS	Betamax	V-2000
Austria	30	10	60
Belgium	70	25	5
Denmark	64	26	10
Finland	70	20	5
France	81	13	6
Ireland	75	15	10
Italy	60	25	15
Netherlands	55	25	20
Norway	80	13	7
Portugal	—	—	—
Spain	40	50	10
Sweden	78	8	14
Switzerland	70	10	20
United Kingdom	70	22	8
West Germany	48	18	34
Total Europe	66%	23%	11%

* Derived from figures given by the Economist Intelligence Unit.[4]

(20%). In the remainder of the countries, the V-2000 reaches only a 9% average, with a total European penetration level of 11%.

Both the Beta and V-2000 formats have been losing ground in Europe, particularly in West Germany, which is Philips' most important market. VHS has received a recent boost as Toshiba, a major Beta supplier, has begun to manufacture VHS units.

Even Philips, despite receiving protectionary measures from the EEC Trade and Industry Directorate, has decided to manufacture a VHS recorder, although it is intended for distribution outside Europe. Many people involved in the industry believe that it is only a matter of time before the V-2000 becomes extinct.

7.4 Software

Regardless of the hardware availability, VCR purchase is due primarily to the demand for the software capabilities that it provides.

The VCR software market has become a multi-million dollar industry on its own, in two major divisions: unrecorded, blank tapes and pre-recorded tapes. Here, the Japanese are far less dominant than they are in hardware. They supply 60 to 70% of the blank tape to Europe, but the Americans and Europeans are largely responsible for the pre-recorded tape market, which is the largest money-maker.

The United Kingdom, France, West Germany, and the Netherlands have become important industrial centres for programme recording, while the US entertainment companies operate on a pan-European basis through European distributors.

7.4.1 BLANK TAPES

For the most part, the video tape manufacturers grew out of the companies that already manufactured audio tapes. Philips and Grundig manufacture tapes for the V-2000, and BASF, Agfa-Gaevert, and 3M manufacture tape in all three formats. BASF tapes are produced in West Germany, Agfa tapes in Belgium, France, and West Germany, and 3M tapes in Wales.

The Japanese manufacturers have attempted to increase their production involvement in Europe in an effort to improve their total market share. Their major manufacturing units are:

Hitachi, Maxell—United Kingdom
Sony—France, West Germany
AEG, Telefunken/JVC—West Germany, United Kingdom

Where higher incomes exist, as in Switzerland, Denmark, and Sweden, sales of blank tapes are quite high. In countries where incomes are relatively low, such as the United Kingdom and Ireland, blank tape sales are significantly decreased.

7.4.2 PRE-RECORDED TAPES

The main sources of programming for pre-recorded tapes are feature films, distributed through the European subsidiaries of American entertainment companies: Columbia International Corporation, Disney, MGM, RCA, and Warner. Thorn EMI in the United Kingdom, RCV Mondiales in France, and PolyGram in West Germany and the Netherlands are major buyers of feature film rights for home distribution. They then record and distribute the videos in addition to their own video output. Their dominance is clearly demonstrated in Table 7.3.

West Germany and the United Kingdom are the two most important markets for video production, and the United Kingdom is the largest and

Table 7.3 Pre-recorded tape distribution[4]

Austria	Warner, Thorn EMI
Denmark	PolyGram, Metronome-Warner, AB Collection, Select Video, Esselte
Finland	Warner, CBS–MGM, Disney, Cinema International Corporation (CIC), Home Video
France	RCA, Warner-Filipacci, Regie Cassette Video, PolyGram, Thorn EMI, Disney, Videorama
Ireland	WEA, Thorn EMI
Italy	Italian Domovideo, WEA Warner, Walt Disney, VCR
Norway	PolyGram, VCL, Thorn EMI, Esselte, RCA, MGM, 20th Century Fox, Mayco, Club Consult, Production Novio
Spain	Warner, Video España, Tele-Tector, Video Technics Ltd
Sweden	Esselte Video, Hemvideofilm, Sonet Video, Polar, Mariann, Scan Video, Thorn EMI, PolyGram, WEA Metronome (Warner)
Switzerland	Videophon (Warner/VA, Columbia, Selectavision), Parvideo (Disney), Tradex (Thorn EMI), CBS Switzerland (VCI)
United Kingdom	Warner, CIC, Disney, Thorn EMI Home Video Holdings, Guild Home Video (Esselte), BBC, Palace Video, Virgin Video, PolyGram
West Germany	Warner, RCA Columbia, Fox, CIC, Taurus, Euro Video (MGM/VA, Disney), Thorn EMI, ITT, Atlal Videothek, Arcade, Constantin Video PolyGram

most competitive. France has also been developing its own industry, but low penetration levels of VCRs have restricted the market somewhat. The Netherlands has the dubious distinction of being the major producer of video pornography, which is a popular programme category.

Over 80% of pre-recorded tapes are rented, rather than purchased. There is a trend towards increased sales brought about by price reductions, more efficient duplications, and strong competition. In the United Kingdom and West Germany, for example, prices have fallen under $20 for a quality pre-recorded education or entertainment tape. In other countries, prices have remained on average proportionately higher.

Rental outlets for pre-recorded tapes have appeared throughout Europe. In West Germany and the United Kingdom, a surplus of rental stores exists, and hard competition and price cuts have removed practically any profitability. Service stations and supermarkets are now carrying limited stocks of recent movie releases. Growth in other countries has not been quite so rampant.

The major pan-European distributors, like Warner, Thorn EMI, RCA, and PolyGram, have tailored marketing strategies to individual countries and the release pattern differs among the countries of Europe. The distribution method in each country has greatly affected the software uptake. In the United Kingdom and Ireland, where many pre-recorded tapes are rented, consumers are of a lower market profile. Where VCRs are a more middle-class item, as in Austria, Switzerland, the Netherlands, and Scandinavia, the purchase of ancillary equipment like cameras, blank tapes, and pre-recorded tapes is much greater. Poor VCR distribution in Italy, Spain, and Southern France has led to limited tape purchase, or even rental.

7.5 Hardware

Since 1979, the three major technical formats have remained the same, but modifications to the machines have been introduced to increase competitive advantages. Fast forward, fast reverse, freeze frame, slow motion, and timing device alterations to enable flexibility in pre-recording have all been added. Current trends are towards miniaturization, and all three format manufacturers have made improvements in their models.

7.5.1 MANUFACTURING

The availability of particular brands of hardware and the presence of specific manufacturers in each country has had varying effects upon VCR production and sales. The Japanese machines are the most popular, but relatively few are assembled in Europe. The top nine video consumer electronics companies in Europe include:[5]

European AEG-Telefunken, Blaupunkt (Bosch), Grundig, Philips, Thorn EMI
Japanese Hitachi, Matsushita, Sony
United States ITT

Most often, manufacturing plants are located in West Germany or the United Kingdom.

Manufacture of the V-2000 occurs in the Dutch and German areas that are closely affiliated with Philips–Grundig: Austria, Belgium, West Germany, and the Netherlands.

The Japanese companies have set up very few manufacturing Euro-bases. The small share of VCRs completed in Europe are assembled from Japanese-manufactured parts. Increased pressure from the EEC has led the Japanese manufacturers to supplement their domestic operations with Euro-bases. VCR manufacturing facilities are now planned for the United Kingdom, France, West Germany, and Austria, the same countries where European manufacture is strongest. Manufacturing units by country are as shown in Table 7.4.

VHS manufacturers predominate in Europe. Many of the domestic brands are manufactured through cooperation with other companies. While they do not achieve a particularly large market share, they add further competition

Table 7.4 VCRs manufactured in each country[6]

Manufacture by country	
Austria	*Sweden*
Philips (V-2000)	Luxor (VHS)
Belgium	*United Kingdom*
Philips (V-2000)	JVC-Thorn (VHS)
	Toshiba (VHS)
Denmark	Sanyo (Beta)
Bang and Olufsen	Sharp (Wales) (VHS)
(assemble parts from Philips/Grundig	Mitsubishi (Scotland) (VHS)
V-2000 in Austria)	GEC (VHS)
France	*West Germany*
Thomson-Brandt (V-2000)	Matsushita-Bosch (VHS)
Philips (V-2000)	ITT-Sel (VHS)
Akai (VHS)	Sanyo (VHS)
Sony (Beta)	Sony-Weba (Beta)
	Hitachi (VHS)
Netherlands	Grundig (V-2000)
Philips (V-2000)	JVC-Telefunken (VHS)

Table 7.5 VCR European manufacturers by format[7]

(a) *Major manufacturers*		
VHS	*Beta*	*V-2000*
AEG-Telefunken (JVC)	Sony	Philips
Hitachi		Grundig
Bosch		ITT (variations
Matsushita		on Philips'
Thomson		machines)
Thorn EMI		
ITT (JVC make)		

(b) *Some others*		
Ferguson	General	Pye (variations
Fidelity	NEC	on Philips'
Sanyo	Sanyo	machines)
GEC	Fischer	Bang and
ITT (JVC make)		Olufsen
Luxor		
Mitsubishi		
Panasonic		
Sharp		
Toshiba		
Akai		

to the already competitive market-place. However, it is apparent that very few have chosen to align themselves with the V-2000 format. The formats and their major manufactured brands are detailed in Table 7.5.

7.5.2 DISTRIBUTION OF HARDWARE

Retail distribution of VCRs varies among European countries and has influenced the type of VCR penetration—rental or purchase (Table 7.6).

In the large countries, VCRs are distributed primarily through department stores, discount stores, and chain stores, as well as specialist video stores. Where electrical appliance stores exist in abundance, such as in the United Kingdom and West Germany, sales have been the highest, and retail competition in West Germany is the strongest in Europe. Although the population of France is notoriously technology minded, its electronic appliance outlets are limited, and the VCR penetration is also relatively low. Countries such as Italy and Spain have poor regional distribution outside a few major cities, and VCR availability is limited and sales levels are low.

The Scandinavian countries have audio, television, and video retailers in place. However, their small population and competitive approaches to retail militates against the established use of discount stores for price reductions through volume sales.

In the countries where Philips has a strong foothold, there are ready-

Table 7.6 VCR distribution by type[4]

Primarily rental
United Kingdom
Ireland
Rental and purchase
Belgium
Denmark
Netherlands
Switzerland
Primarily purchase
Austria
Finland
France
Greece
Italy
Norway
Portugal
Spain
West Germany

made outlets in place for its products. V-2000 sales have been strongest in these areas.

Throughout Europe, the outlets for VCRs have been growing steadily and many department and general stores have developed special departments to handle video equipment.

In the United Kingdom and Ireland, television rentals is a major industry. Many consumers are already in the habit of renting television sets, and an additional VCR payment per month is quite acceptable. As a result, Ireland has reached penetration levels well above what would be expected with respect to income and wealth of the country.

7.5.3 IMPORTS/EXPORTS

Trade among countries in the EEC is very volatile, and VCRs are no exception. Even the countries which manufacture most of the VCRs rely upon Japan for the bulk of the machines. Imports outweigh exports by anywhere from three to ten times. Although the Japanese have planned to open more Euro-bases, which will lower their imports to Europe, the machines will still be primarily Japanese. This move towards joint-venture plants in Europe was strengthened in July 1985 when the EEC announced that it would raise the tariff on imports of VCRs from 8% to 14%, a move that will decrease the volume of exports from Japan.

The major European manufacturing companies are also the major exporters—West Germany, Austria, the Netherlands, and the United Kingdom. Most of their exports are destined for countries within Europe.

7.6 Key regulatory concerns

Because the Americans were relatively late to enter the VCR market and have maintained a low profile in Europe and because the Philips–Grundig V-2000 did not capture a significant market share, the Japanese manufacturers have experienced overwhelming success. As a result, there has been a strong reaction against them from European manufacturers, with the introduction of trade barriers and protections.

In November of 1982, Philips–Grundig filed complaints with the European Commission, alleging that leading Japanese manufacturers were 'dumping' VCRs. The complaints were aimed primarily at Sony, which was trying to boost the Beta format through price reductions in the lower and middle range models. Grundig responded with its own price cuts and reduced its staff members working on the V-2000.

At the same time, the French government implemented an 'import' policy designed to slow down Japanese VCR entry into France. All foreign VCR shipments were routed through Poitiers, a small, poorly equipped and relatively inaccessible customs post. All documentation had to be completed in perfectly correct French. This caused a large build-up of Japanese VCRs and nearly halted their entry into France completely. Although the V-2000 was also a 'foreign' import, the German and Dutch manufacturers were more familiar with French customs, so their imports were hardly impeded.

In the United Kingdom, the government threatened to impose restrictions on Japanese imports, while examining paths of entry for Japanese investment to create more jobs.

The negative backlash from the major European industrial companies resulted in negotiations between the Japanese Ministry of International Trade and Industry (MITI) and the EEC Trade and Industry Directorate. Following a month of negotiations in February 1983, the Japanese agreed for the first time to limit exports of a particular item to the EEC. All previous trade agreements had been between Japan and individual countries. A 3-year agreement limiting Japanese VCR exports to the EEC was put into place, as well as guidelines to regulate VCR prices.

The agreement limits Japanese VCR exports to 4.55 million units per year, including 600 000 units to be constructed in Europe. In addition, the agreement provides for MITI to establish a floor price for the exports. It is designed to align retail prices of the Japanese units with those produced in Europe.

The agreement includes a price agreement clause which allows Japanese ex-stock and European ex-factory prices to be aligned and protects Philips and Grundig from unfair price competition. As a result, Philips and Grundig withdrew their complaints against the Japanese; the French, however, continued to persist in their Poitiers slowdown.

The benefit of this move for Philips and Grundig, and the V-2000 format, is that it allows them to reach production levels to compete in the world market, without serious threat of price competition.

Public and government criticism was levelled against this policy for several reasons. First, it is apparent that Europeans prefer Japanese consumer electronics products in general. Second, protection of the Grundig–Philips V-2000 does not guarantee that it will be competitive world-wide; only consumer acceptance can do that in the absence of an industry standard. Historical evidence suggests that consumers are not likely to change their preference patterns. Third, the price freeze on VCRs at the level of ex-factory performance still gives Japanese manufacturers 'good profits'. Fourth, because the VCR restrictions include kits to be assembled in Europe, there is very little incentive for the Japanese to open more Eurobases for VCR production, an issue of particular concern to the British and West German governments. Finally, VCR demand in Europe is slackening off, and the restrictions will probably not impair the relatively dominant Japanese position.

The MITI has already recognized this fact, and has requested that Japanese manufacturers cut down on these exports to the EEC at least by 10%, in order to maintain price levels and avoid competition among the Japanese manufacturers. As a result, the Japanese will continue to dominate the European market, with the VHS and Beta formats remaining strong.

Video piracy is a thriving industry in Europe, which is at its worst in the United Kingdom and the Netherlands. Illegal tapes, particularly of films, are produced and then offered at prices which undercut those distributed legitimately. Estimates of illegal units in circulation range from 50% in Germany to 70% in the Netherlands and over 75% in the United Kingdom, where the rental market is the strongest. The United Kingdom has become the pirate centre of Europe because it is a large market and competition is stiff. While the United Kingdom suffers from outright copying, piracy takes other forms elsewhere. France has encountered a problem with copyright infringement and, along with Italy, it suffers from theft of the tapes themselves.

Most countries have a particular 'window' or time frame between release of a film at the cinema and release on the home video market. This is to prevent further competition for the cinema which has been undergoing a general decline in Europe (Chapter 4). Contraband videos put many films on offer while they are still at the cinema. Researchers have discovered that video release near the peak of cinema attendance of a film can enhance video and rental sales. Pirated copies, however, preclude film producers from realizing this profit. As a result, piracy could ultimately affect the budgets of major productions adversely.

7.7 Video disc players

While VCRs have experienced considerable success in Europe, as a consumer product, video disc players (VDP) have not. The lack of software and inability to compete as a viable alternative to VCRs have provided major obstacles to the acceptance of video discs.

Philips and Drew forecast 350 000 units in annual sales by 1985 cumulating to a total market of 520 000 units. A revised forecast from the *Economist* Information Unit predicted a total market of 350 000 units by 1985. However, 1983 sales at fewer than 500 units per month across Europe indicated that VDPs were unlikely to reach even half the anticipated consumer market.

7.7.1 FORMAT

Three VDP formats are available in Europe, two of which are for home consumers. At present, the United Kingdom is the only country in which all three formats are available. Philips' Laservision is a laser optical system which is the most expensive, most sophisticated, with good picture quality and interactive capabilities. Millions of microscopic pits are etched into the smooth plastic disc. These pits deflect the laser beam to produce information which is translated into programme signals. These discs are virtually immune from scratches and fingerprints.

RCA's Capacitance Electronic Disc (CED) is the cheapest and least sophisticated format. The disc is similar to a phonograph record and is subject to wear and tear. The information is etched into a single fine groove and is tracked by a stylus with an electrode and then translated into the programme signal. Much like a phonograph record, the disc is loaded into the player from a special protective holder and the signal can be obstructed by dirt, dust, or fingerprints.

The medium-range model, JVC's VHD, has never been introduced to home users and is available only for industrial and educational use. While it is also a capacitance model, the VHD system is more sophisticated than the CED and has interactive capabilities. The disc has three tracking grooves, but only the centre groove contains any information. The tracking guides keep the stylus centred over the information area but there is no actual contact between the stylus and information groove, so there is no wearing down as with the RCA Capacitance.

7.7.2 VDP INTRODUCTION

VDPs were first launched in Europe in May 1982 in the United Kingdom by Philips. Pioneer launched its own Laservision model in December of the same year. Despite expectations of 100 000 units in sales, only 10 000 units were sold during the first 12 months. Limited availability of software at the launch presented a major sales problem.

In September 1982, Laservision was launched in West Germany. After 6 months of heavy marketing and promotions, only 10 000 units had been sold. Laservision was also launched in France in 1982 with similar limited success.

Despite poor commercial acceptance of Laservision, RCA launched its CED system in the United Kingdom in collaboration with Hitachi and GEC. It focused its marketing on a low price, selecting the United Kingdom as the initial market, believing it had the best opportunity for success in Europe.

Meanwhile, due to poor market performance, Philips postponed its launches in Austria, Switzerland, Sweden, and the Netherlands. Thorn EMI, which had planned to launch JVC's VHD player for consumer use in the United Kingdom also postponed its plans. Thorn EMI later launched the VHD in the fall of 1983 for industrial and educational uses.

VDP price reductions of 25% were introduced in 1983 to increase ability to compete with VCRs. However, they did little to stimulate market growth.

Further problems with distribution arose because no large retail chains sold VDPs, and manufacturers had great difficulty in persuading retailers to stock them. As a result of poor sales, RCA announced that it was withdrawing its CED player world-wide in April 1984, admitting that VDPs had failed to become a viable consumer substitute for VCRs.

Estimates indicate that only about 25 000 VDP units have been sold across Europe, a negligible percentage. While growth in the consumer market is expected to remain very limited for a number of years, the industrial/educational market is expected to exceed the consumer market by 1987.

7.7.3 USERS

VDPs were targeted at the same initial market as VCRs—up-scale, working professionals. However, VDPs have not been viewed as a substitute for VCRs and most VDP users are also VCR owners who view VDPs as a speciality item to supplement their audio/visual systems.

As a result, the consumer market for VDPs in Europe is very limited. Recent trends indicate that future users are likely to be companies and institutions rather than individuals.

7.7.4 SOFTWARE

VDP discs for consumer use are available in both the CED and Laservision formats. They retail at $15 to $20, and most VDP discs are purchased, rather than rented, as there are few rental outlets. In addition to general entertainment and films, many minority-interest discs are manufactured.

The best-selling discs are pop videos, and both Philips and RCA have rapidly expanded their music choices. Philips obtains its software programming from: PolyGram, 20th Century Fox, CIC, Rank, Introvision, Precision, Guild, and the BBC. Together, Philips and Pioneer have about 400 titles available to consumers.

While minority-interest subjects have been targeted as marketable programming for VDP users, even low-budget productions require investments of several thousand dollars. Therefore, the minority-interest discs can be costly to produce.

Philips has a manufacturing plant for discs in Blackburn, Lancashire, in the United Kingdom and Laservision discs are also pressed by Mullard in the United Kingdom and PolyGram in West Germany. Both RCA and Philips are still producing discs for their existing players.

7.7.5 FUTURE APPLICATIONS

Prospects for the consumer VDP market are not promising. The industrial market, however, holds considerable potential. The United Kingdom and West Germany are European leaders in VDP use for advertising, promotion, and staff training.

The optical reflective format is the most preferred by industrial users, so Laservision has good industrial acceptance. It is the most sophisticated and durable, so the discs can be used repeatedly without damage, unlike VCR cassettes, which wear out with use. When this is added to its interactive potential, Laservision becomes particularly well suited for use in on-site sales or training. Retail stores, such as Olympus, Top Shop, and Mothercare and the Post Office in the United Kingdom, are using VDPs for point-of-sale advertising and promotion. Other companies, such as Marconi Space and Defence Systems, British Telecom, CEGB, and American Express have investigated VDPs' interactive uses for in-house applications as well.

IBM Europe also uses VDPs at point-of-purchase for personal computer (PC) training and promotion. Together with Philips and EPIC Industrial Communications, IBM uses discs manufactured by Philips to give instructions and promote PC use in English, French, German, Italian, Spanish, and Dutch. JVC and Thorn EMI have collaborated to launch a VHD VDP for interactive use with home computers.

VDPs have also been used with teletext to supply viewdata services. In the United Kingdom, the English Tourist Board, together with Prestel, has a VDP combined with microcomputer use to provide a picture, while teletext data is superimposed. The teletext can be constantly updated, and an updated viewdata service ensues. The travel and tourism industries have been targeted for such systems.

One of the successful applications of VDPs has been for video jukeboxes. The durability of the discs has made them well suited for this market.

7.7.6 SUMMARY

In the immediate future, the home VDP market is likely to remain small, with little growth. Competition with VCRs has proved to be difficult, and the inability to record programming has made VDPs less desirable than VCRs to consumers. As a result, VCRs have taken the overwhelming share of the video market.

However, for industrial use, VDPs appear to have advantages over VCRs, particularly durability with repeated use. Philips Laservision is the preferred format for industrial use. As the dominant VDP manufacturers in Europe, it is likely that it will continue to shift its emphasis from the flagging consumer market to the more promising area of industry.

7.8 Compact discs

While VDPs have not been successful in the consumer market in Europe, compact discs, a related audio digital technology, have experienced rapid consumer acceptance.

7.8.1 BACKGROUND

The technology for compact disc players (CDP) was developed by Philips in the seventies; in 1979, Sony and Philips collaborated to develop the final product, which was first introduced in 1981. Eight CDPs were introduced into Europe in 1983, manufactured by Philips, Sony, Marantz, Hitachi, Technics, Toshiba, Fisher, and Akai. At the same time the compact discs (CD) for the players were also launched by PolyGram in West Germany, the Netherlands, France, and the United Kingdom. While the CDPs and VDPs have been compared because their technology is similar, the firm commitment to software by CD producers and manufacturers, and simultaneous software and hardware release, gave CDPs a distinct advantage that VDPs never had. In addition, there are many more hi-fi enthusiasts than video enthusiasts, providing a greater target market for the product.

In 1983, 160 000 units were sold in Europe, UK sales being approximately 30 000 units, West German sales 35 000 units, and French sales 20 000 units. These sales to reached 220 000 units by the end of 1984 and are expected to reach 2.7 million units by 1986, reaching 25 to 30% of the audio market. The United Kingdom, West Germany, Sweden, and the Netherlands are expected to be the dominant markets. With the addition of in-car and personal use, Euromonitor predicts CDPs will become the dominant audio medium by the end of the decade, reaching $1.7 billion in sales.[8]

7.8.2 MANUFACTURE

A number of European companies are beginning to manufacture CDPs, and retailers are stocking them as a standard item. When Sony, Hitachi, Akai, and Panasonic adopted Philips' system, it became the established standard, allowing manufacturers to proceed without the problems of format selection, which has plagued both the VCR and VDP markets. There are presently over 38 European manufacturers licensed to produce CDP systems.

Hardware costs vary slightly from country to country and the two major markets of the United Kingdom and West Germany listed CDP prices at $750 to $900 respectively in 1984. Sony supplies the top end model, while Philips supplies the low end, targeting the mass audio market.

CDPs with amplifiers and tuners used as an alternative rather than supplement to regular stereo equipment were brought on to the market by the end of 1984. In-car and personal CDPs have also been launched. The main manufacturers with models available in Europe include:[9]

Akai	Pioneer
Dual	Revox
Ferguson	Rotel

Fisher	Sansui
Hitachi	Sentra
JVC	Sony
Marantz	TEAC
Meridian	Technics
Mission	Toshiba
NAD	Trio
Philips	Yamako

7.8.3 USE

Most CDP owners are hi-fi enthusiasts who are eager to obtain the latest state-of-the-art audio equipment. However, CDPs are fast becoming a mass consumer item, as indicated by increasing public awareness and inclination to purchase—up to 50% with intention to buy in 1984 from 15% in 1983.

7.8.4 SOFTWARE

The growth in the CDP market was stimulated by the availability of software at the time of launch and the increasing number of titles added since that time. The majority of discs pressed for the European market are made in Japan and West Germany.

PolyGram, the record company jointly owned by Siemens and Philips, is the main European CD producer. Its plant in Hanover, West Germany, began production in August 1982 and, by June 1984, had pressed its 10-millionth disc. The initial phase to make the plant operational cost $8 million and an additional $12 million was necessary to begin production. Another $12 million has been added to expand the factory because its current production level cannot fulfil the demand. PolyGram's production level is 50 000 discs per day, rising to 80 000 by the end of 1984. By 1985 annual production had risen to 14 to 15 million discs and 1000 titles.

PolyGram is best known for its Philips and Deutsche Grammophone labels, and presses discs for Sony and Warner Communications. CBS, Arista/Ariola, RCA, Chrysalis, Virgin and WEA are also anxious to begin CD production.

Twenty to thirty other manufacturers plan to produce CDs. Nimbus in Wales has been producing 50 000 CDs per month. Philips' Laservision plant in Blackburn, the United Kingdom, plans to convert its video disc operation to produce CDs and Thorn EMI's video disc factory in Swindon is primed to be converted to CD production as well. EMI is also involved with Toshiba to produce 2 million discs per year in Japan.

The average cost for discs in Europe is $14 for pop and $18 for classics. The average user purchases 25 discs, while only 15 had been forecast. At this rate, it is estimated that over 5 million discs had been sold throughout

Europe by 1984. By 1989, disc sales are expected to reach $1.6 billion annually. Excellent quality control has ensured that most discs are usable, which has improved cost efficiency.

7.9 Conclusion

Growth forecasts for Europe indicate that VCRs will continue to hold a very large market potential over the next 5 years (Table 7.7). While the leading markets of the United Kingdom and West Germany have experienced a reduction in growth, other countries such as France, the Netherlands, and Norway should compensate with high growth, reaching a high VCR penetration level by the end of the decade. The current leaders in VCR take-up, such as the United Kingdom, Sweden, and Germany will continue to remain in the forefront.

The countries which are weakest economically, such as Portugal, Spain,

Table 7.7   VCR growth in Europe of estimated VCR sales

(a)	*Market value — Western Europe* ($ billion)	
	1983 4.1	
	1984 4.4	
	1987 4.8	
(b)	*Growth forecasts*	
	High	*Low*
	France	Spain
	Netherlands	Sweden
	Ireland	
	Norway	
	West Germany	

and Italy, are not expected to grow significantly or rise above a low penetration level.

VHS has proved to be overwhelmingly the most popular format. The V-2000 has its greatest reach in the German-speaking countries, but it is likely to be phased out, while Beta should retain a substantial market share; countries which are expected to experience high growth are more favourable towards VHS. As a result, the VCR market, which should exceed $5 billion by 1990 is likely to be dominated by VHS manufacturers.

The home VDP market is not likely to experience much growth. Inability to record and competition from VCRs has made VDPs the loser in the home video market. At the same time, the durability of the video discs themselves and their ability to hold large amounts of information will encourage their use in the industrial sector.

CDs are likely to grow in use during the last half of this decade. They are continuing to increase in popularity and are predicted to become the dominant audio medium by the 1990s. They are perceived as strictly an audio component and represent no real competition with VCRs.

It is apparent that VCRs are currently the dominant video medium in Europe and their continued growth ensures that, at least for the next few years, they will remain so. The recent trends in VCR growth decline in the United Kingdom and West Germany may be repeated throughout Europe and the upper limit to VCR penetration in Europe may be 50%, rather than the 80 to 100% originally predicted.

References
1. 'VCR: world turns on', *Broadcast*.
2. Ogilvy and Mather Europe, *The New Media Review 1984*, Ogilvy and Mather Europe, London, 1984.
3. 'Telefirst—a big tele-failure', *CableVision*, 25 June 1984, p. 14.
4. *Economist* Intelligence Unit, *The Home Video Revolution in West Europe*, *Economist* Intelligence Unit Ltd, London, 1983.
5. Mackintosh Consultants Ltd, *Mackintosh Yearbook of West European Electronics Data 1984*, Mackintosh Consultants Ltd, London, 1984.
6. 'A sales boom for VCRs', *Financial Times*, 18 July 1984, p. 16.
7. Mackintosh Consultants Ltd, 'European market leaders', *European Electronics Companies 1984/85*, Benn Electronics Publications Ltd, Luton, 1984.
8. Euromonitor Publications Ltd, *Compact Discs*, Euromonitor Publications Ltd, London, 1984.
9. *Home Electronics*, various issues.

8

Entertainment services: Plans and opportunities for cable and satellite

8.1 Introduction

The number of national channels proposed for cable and DBS by 1990 is almost double the number of national broadcast networks in place in 1985. This does not include the myriad of local and regional programmes proposed by local groups and individual cable operators.

The implications are not trivial. For example, if we assume that each of these channels actually provides an average of 8 hours of programming per day, then the annual programming requirement, some 120 000 hours, exceeds, by a factor of around 250 times, the total amount of film product coming out of Hollywood each year. That is, the programming demands constitute over 80 000 full-length 90-minute feature films or the equivalent in series, sports, news, music, live, and other programming.

Naturally, some projected channels will fail to materialize and some will involve little or no new product at all. There will also be 'wildcards' emerging between now and 1990 which are not included below.

The early trend in Europe, however, is clear:

More entertainment Social, cultural, educational, and political objectives determine the programming philosophy in most national broadcast systems. The so-called quality of programming may be high but the existing channels are not programmed as family entertainment channels.

More movies Historically, consumers have paid to visit the cinema and more recently to hire (or purchase) video cassettes. The cinema is in decline (Chapter 4), the VCR market is thriving (Chapter 7), and the cost of film rights, with the often jealously guarded position of the cinema in the release schedule, has meant a 2- to 3-year delay in movies appearing on television.

Emergence of narrowcasting Once the cable or DBS infrastructure is in place it becomes feasible to offer programming for minority- or special-interest tastes. Europe began its narrowcasting with music, sports, and children's programming.

In the United States, for example, there were ethnic and religious group channels before a sports channel, children's channel, news channel, or music channel emerged (in that order from 1979 to 1981).

More of the same The existing broadcasters are responding to the challenge by offering minor variations on their existing themes—programming to satisfy all cultural and political tastes—by repackaging and retransmitting their broadcast programmes.

8.2 Early pay-television

Prior to the launch of the European Communications Satellite Eutelsat 1-F1 on 16 July 1983, non-national broadcasting was extremely limited. Italy, with its 500 terrestrial television stations was a unique exception. What little cable existed was primarily for must-carry channels and included a small number of foreign spillover services, notably Radio Télé-Luxembourg (RTL) and Télé-Monte Carlo. In addition, pirate broadcasters in the Netherlands were able to tap illegally the NOS transmitters after the close of programming. For some time, they could transmit pornographic and other material to very limited geographical regions of Amsterdam.

Pay-television, too, was limited and available in only four countries: Finland, Switzerland, the United Kingdom, and France. Since 1978, Helsinki Cable Television (HCTV) in Finland has offered a pay-television channel, The Entertainment Channel. It consists of old movies, series, and sports. Of the current 101 000 homes connected to the cable system, over 23 000 subscribe to the pay service.[1] HCTV proposes to upgrade this service in 1986 to a super pay service featuring newer theatrical releases.

In Switzerland, a pay-channel was offered on Zurich cable by Teleclub (51% equity by Rediffusion and 49% by Beta-Taurus). It cost $10 per month and offered a selection of eight old and relatively new movies each month plus cable news information.

In the United Kingdom, the Home Office approved 11 pilot pay-television schemes in May 1981. These were single-channel services on the old four-channel and six-channel twisted-pair cable systems and have since ceased transmission or been incorporated within new cable systems. There were four services: Starview (Rediffusion), Cinemotel (Thorn EMI), Selec TV (Selec TV), and Showcable (BBC Enterprises). Take-up rates for the basic service were generally low (20 to 30%) and the pay service penetration even lower (2 to 10%). The average monthly fees charged by cable operators were from $10 to $13.

In France, there was no pay-television except for 'Video Cinema', a service run by Hachette on a privately owned estate in the Defence quarter of Paris. The service cost $16 per month and offered eight recent films.

8.3   Pioneer satellite-delivered channels

On the Eutelsat Orbital Test Satellite (OTS-2), launched in 1978, two television services were tested. In April, Satellite Television plc began transmitting 2 hours of advertiser-supported English language programming to cable networks in Malta, Finland, and Norway.

During the following month, the European Broadcasting Union (EBU) conducted a 5-week test transmission called Eurikon to 'look into the problems of content, style, organisation, coordination and financial and legal matters which would be raised by a European television programme'.[2] The trial consisted of 35 hours of encrypted broadcast material sent for closed-circuit use to the participating broadcasting organizations—Algeria, Austria, Belgium, Finland, France, West Germany, Greece, Ireland, Italy, Malta, the Netherlands, Norway, Portugal, Spain, Sweden, Switzerland, Tunisia, the United Kingdom, and Yugoslavia.

The replacement satellite, Eutelsat 1-F1, provided the first real opportunity for European-wide television programming as well as the first opportunity for any cable system within a country to receive a non-'broadcast' channel. Although the national allocations of transponders were made by Eutelsat to the PTTs in 1982, the PTTs in most instances were slow to reallocate transponders to lessees. They also found it difficult and expensive to construct and install uplink facilities.

Lessees in turn found that national governments had not formulated policies on private television (West Germany), pay-television (the Netherlands, West Germany, Belgium), or advertiser-delivered programming (Belgium). For example, at the end of 1984, three Socialist Democrat Länder in Germany, North Rhine-Westphalia, Bremen and Hamburg, did not permit private television, even the newly developed German service SAT-1. The Dutch and Belgian transponders had not been used commercially. Euro-TV, the original lessee in the Netherlands, had relinquished its lease to the NOS/EBU consortium later to be called Olympus, and still later, under a threat of court action by Olympus cameras, Europa-TV. Teleclub did not have the right to transmit in its key language market, Germany, and even in the United Kingdom, the contract for the second transponder on Eutelsat 1-F1 between British Telecom and Thorn EMI was not signed until the end of June 1984, about 12 months after the satellite was launched.

The limited national allocations of transponders on Eutelsat 1-F1 and Eutelsat's stated objective that Eutelsat 1-F2 (launched August 1984) was to be primarily for telecommunication services left a short-term unfulfilled demand in the transponder market for all the national groups wishing to capitalize on the proposed development of national cable programmes. To supplement this shortfall, British Telecom in the United Kingdom took an option to lease up to six transponders on an Intelsat VA-F10 satellite (27.5 degrees west). By switching the beams to focus on the United Kingdom, it was possible to achieve a reasonably effective isotropic radiated power (EIRP), a relative measure of the beam across its target coverage, using a half-transponder.

	Beam centre	
	West beam	East beam
Full-transponder	46 dBW	45 dBW
Half-transponder	43.7 dBW	40.7 dBW

The Deutsche Bundespost adopted a similar situation by leasing up to six transponders on an Intelsat V Indian Ocean Spare for programmes in West Germany until Kopernicus, the communications satellite, or TV-SAT, the DBS satellite, begin operation. The French telecommunications satellite Telecom 1 is being used in a similar manner.

The United Kingdom was the first country to begin developing additional satellite-delivered channels totally outside the existing broadcasting infrastructure. By the end of 1985 there were eight new programming services in English available to cable households in the United Kingdom and, in some cases, Europe:

General entertainment	Sky Channel
Sport	ScreenSport
Popular music	Music Box
Children's programming	The Children's Channel
Movie channels (2)	Mirrorvision
	Premiere
Targeted	Lifestyle
	The Arts Channel

The following sections detail the new channels that were operational by end of 1985 and then consider the picture for the next couple of years as it is emerging. As in the United States, there are many plans, some of which are well publicized, fewer services that actually make it on air, and fewer still that have the staying power to survive in the new markets. This chapter necessarily provides a snapshot of Europe as well as giving an indication of the direction of the new programme services. The channels are divided broadly into European-wide and national or regional services and then again into operational and planned. Each channel is discussed in terms of its programming, viewer market, and potential.

8.4 Pan-European services

Pan-European services are defined as those channels which are being distributed by satellite and received in at least three European countries or which are proposed for either European-wide or multi-national distribution. There is an element of arbitrariness in the definitions, as in some cases we have classified channels on the basis of stated *intentions* for the service.[3] The operational services intended for Europe are:

1. General and entertainment services
   (a) Sky Channel
   (b) RTL-Plus
   (c) Europa
   (d) SAT-1
2. Rebroadcast services
   (a) TV-5
   (b) 3-SAT
   (c) RAI (relay)
3. Specialized or narrowcast services
   (a) Music Box
   (b) New World Channel
   (c) World Public News
   (d) Cable News Network
   (e) ScreenSport

8.4.1 GENERAL ENTERTAINMENT

Sky Channel

Sky Channel is located on Eutalsat 1-F1, transponder 6. The transponder is eclipse-protected and the service has been encrypted. News international is the majority shareholder (90%) in the advertiser-supported service with D. C. Thompson, Guiness, Mahon, Ladbrokes and institutional shareholders taking the remaining 10%.

Sky Channel offers a mix of music, soap operas and dramas, sport, and movies—a classic entertainment channel in the US sense. It began with 2 hours of programming per day on OTS in April 1982, increasing to 5 hours with the switchover to Eutelsat. By February 1985, it offered 73 hours per week of programming, and by October this had risen to 112 hours. It is positioning itself as the family entertainment channel of Europe with particular emphasis on the 16 to 34 year old group. A modular programming schedule attempts to match content with target groups throughout the evening—music for the younger age groups, followed by family entertainment and more adult-oriented series, movies, and dramas. The evening concludes with sport and late night music and movies.

Local programming comprises the music segment called Sky Trax, a weekly interview/music show, and, commencing in mid 1985, a segment of locally/jointly produced children's programmes. The music segment comprises of three 1-hour shows per weeknight, hosted by disc jockeys, plus a European Music Chart show on Sundays. In addition Sky began to expand its programming hours in mid 1985 by introducing 4 hours of children's programming, *Fun Factory*, on weekend mornings, repeating its music programmes on weekday mornings and adding extra weekly series ('strip' programming). The result was up to 17 hours of programming per day by late 1985. The remainder of the programming is US, Canadian, Australian, and European series and entertainment. Programming costs are estimated to be around $1000 to $2000 per hour for bought-in programming, and considerably more for locally produced material.

Sky Channel is advancing across Europe, slowed only by the tardiness of national governments to authorize downlinks for foreign-originated advertiser-supported programming and the lack of cable systems throughout Europe to carry the services. The next key opportunity for Sky is to penetrate the various national 'SMATV' markets of Europe as well as the consolidation of the cable networks particularly Belgium where there are 2.8 million cabled households. Sky Channel is the pioneer and the largest of the new channels. It achieved a penetration of just under 3 million households by early 1985 and forecast 8.6 million by 1987.

As of August 1985, Sky Channel's estimated household penetration was:

Netherlands	2 042 000
Switzerland	653 000
Norway	136 000
Finland	152 000
Austria	137 000
United Kingdom	124 000
West Germany	491 000
Belgium	260 000
Sweden	76 000
France	3 000
Denmark	1 000
Luxembourg	4 000
	4 079 000

Note: in France and Denmark the figures refer to hotel rooms, not residences.

Sky Channel generates its revenue through advertising and through sales of its programmes throughout the world. It is pioneering pan-European product advertising. This is in sharp contrast to similar stations in the United States who charge cable operators a small fee based on the number of households receiving the service. Sky Channel follows the Independent Broadcasting Authority's guidelines for television advertisers in the United Kingdom and offers up to 6 minutes of advertisements per hour. Additional constraints required by downlinking countries limit the range of products to be advertised (alcohol, tobacco, and pharmaceutical products are excluded) and the placement of advertisements (blocks at the beginning and end of programmes or at junction breaks in the programme).

In early 1985, Sky Channel's rate card of $600 per 30 second spot, combined with a 'good' Sky Channel audience of 3% of those households able to receive Sky Channel, computes into an average network CPT of around $5.50—a cheap buy on European television. As Sky Channel's penetration is increasing monthly, at least for the next couple of years, the CPT rate will improve 'automatically', and the total size of the audience delivered by Sky Channel will approach network levels of 300 000 to 600 000 per peak-time programme. Regular adjustments in the price appear necessary for growth reasons alone. In May 1985 the advertising rates doubled, so that a prime-time 30-second spot cost $1200 and in September 1985 the rate almost doubled again to $2100.

Sky Channel is the first real television alternative to traditional broadcast television advertising in Europe. By maximizing its overall market position it offers European-wide and in some cases national advertising opportunities.

RTL-Plus

RTL-Plus is a German language entertainment channel originating in Luxembourg by the Luxembourg private broadcasting organization CLT (Compagnie Luxembourgeoise de Télévision). It began satellite transmission in August 1985 to over 250 000 cable households in Europe from transponder 8 on Eutelsat 1-F1 (an east spot beam which was designated originally as a spare). The equity in the channel is 60% CLT and 40% Bertelsman with the major shareholders in CLT being Audiofina (54.5%), Schlumberger (12.6%), Hachette (8.2%), Paribas (10%) and Rothschild (5%). The German newspaper group WAZ (Westdeutsche Allgemeine Zeitung) is also negotiating an equity position in RTL-Plus.

The CLT/Bertelsman's German-language entertainment channel began in January 1984 by broadcasting terrestrially as a 'spillover' service for 6 hours each evening from 17.30 until 01.30 in the Saarland and Rhineland-Palatinate areas. It has now added the service via Eutelsat which comprises: general entertainment programmes, including drama series (30 to 40%); films; local, national and international news; music; and games shows. Much of the programming comes from the United States. Programme costs amount to about $2000 to $3000 per hour, and up to $4000 per hour is budgeted for special programming. In addition, some programming is generated at RTL's studios in Luxembourg.

The CLT/Bertelsman terrestrial channel reached 900 000 viewers from a 1.3 to 1.7 million potential in Saarland, and Rhineland-Palatinate areas in West Germany. Over half of the potential viewers have at least considered taking the channel. A special aerial is needed to receive the programmes for the terrestrial subscribers. The Eutelsat transponder is only a temporary location for the channel. There are plans to take over the Italian west spot beam when the Olympus satellite is launched. RTL negotiated a position with the French to take two transponders on TDF one for RTL-Plus, a deal that was prepared before the Bredin Report and subject to uncertainty. By September 1985 the use of TV-SAT, the German DBS satellite, had also become a serious option for RTL-Plus.

An average of over 10 minutes of advertising per hour supplies the revenue for RTL/Bertelsman. The advertising adheres to pan-European guidelines, forbidding children's advertising, tobacco products, and alcohol. Their yearly budget is approximately $6 million. They expect to have a deficit of $2.2 million in the first year and reach profitability by February 1987.

European Programme on Satellite-Europa
The European Programme on Satellite will be distributed via Eutelsat transponder 3, the Dutch transponder, until the launch of the European Space Agency's satellite Olympus. Equity in the channel was shared among several national broadcasters, the NOS (Netherlands), ARD (West Germany), RTE (Ireland), RAI (Italy) and RTP (Portugal), with the possible addition of Spain, and is sanctioned by the EBU.

The EPS evolved from the EBU's pan-European news experiment Eurikon, and is based at the NOS in the Netherlands. The service will be a multi-language (using four audio channels), pan-European channel broadcast at 19.00 to 23.00 daily, building eventually to 10 hours daily. The evening block will be repeated the following day at 09.00 to 13.00. Programming will consist of children's programming, a half-hour newscast, documentaries and films with programming divided into different themes for each evening of the week. The channel hopes to produce 20% of the programming. The daily newscast will be provided by an international television news agency.

All programming will be in the original language, but some translations will be possible using teletext subtitles and additional audio channels. A maximum of 15 languages will ultimately be possible. Countries involved include Portugal, Spain, Ireland, the United Kingdom, Canada, Belgium, Switzerland, Greece, Italy, West Germany, and Finland. The programmes will come from its own staff, some EBU programmes, some Eurovision programmes, and some EBU members' archives. Some news bulletins will be contributed by broadcasters from West Germany, Switzerland, Belgium, and Ireland. Cable News Network of the United States also plans to provide news programming as well.

The service is intended for a European-wide audience as the programming is in multiple languages. In the Netherlands, it is a 'must carry' channel for cable operations and it will be relayed terrestially in Portugal and Spain. The target market is comprised of the young, mobile professional audience. The Netherlands Ministry of Welfare donated $4 million and will allocate $2 million as the NOS contribution for start-up costs. Three further annual subsidies of $1.8 to $1.9 million will be given. Additional revenue for the channel will be generated through advertising.

EBU guidelines for advertising on DBS will be followed, with up to 10% of transmission time for advertising.

SAT-1

Sat-1 originates in West Germany and is transmitted on an eclipse-protected transponder on Eutelsat 1-F1, transponder 10. SAT-1 is an advertiser-supported general entertainment channel which is encrypted. Its partners with their respective equity contributions include:

PKS—Programmgesellschaft für Kabel- und Satellitenrundfunk (40%)
APF—Aktuelle Presse Fernsehen (160 newspapers) (20%)
Axel Springer Verlag (9.9%)
Burda Verlag (8.2%)
Kabel Media Programmgesellschaft (6.6%)
Heinrich Bauer Verlag (6.1%)
Euromedia Gesellschaft (Holtzbrinck) (5.4%)
and small shares for:
    FAZ—Frankfurter Allgemeine Zeitung
    Verlagsgruppe George von Holtzbrinck (Euromedia Gesellschaft)
    Wolfgang Fischer Katelmedia Programmgesellschaft
    Neue Mediengesellschaft

SAT-1 began as PKS in January 1984 offering 6 hours of daily programming to cable subscribers in Ludwigshafen. At the beginning of 1985, a reworked consortium channel began broadcasting 12 hours per day via Eutelsat F-1. The programming is German-language general entertainment, comprised primarily of US television series and feature films. Daily programming is provided by: PKS—60%; APF—14% comprising a news slot; Bauer, Springer, Burda, and Holtzbrinck—a combined total of 26%. Programming costs are estimated at over $1.2 million per week.

The audience market for SAT-1 was 250 000 viewers, providing programming to German-speaking audiences on the cable systems in Ludwigshafen and Munich in West Germany and cable systems in Austria and Switzerland as well. Research commissioned by SAT-1 indicates that the average subscriber to the service watches 52 minutes of the channel per day. By the end of 1985 the service was available only in West Germany. It was being received by around 500 000 cable and SMATV subscribers. SAT-1 is advertiser supported, with an average of 12 minutes of advertising per hour, broadcast in blocks after 20.00.

8.4.2 REBROADCAST SERVICES

These are services which comprise repackaged programming from national broadcasters. They constitute a 'cheap' path of entry into this market for broadcasters.

TV-5

TV-5 is a multi-national cultural public channel which originates from France (and is transmitted on Eutelsat 1-F1, transponder 4. It is encrypted using the RTC-Discret system. Equity in the channel is distributed among the three French national broadcasting services TF1 (28.6%), A2 (28.6%), and FR3 (14.3%), the French Belgian broadcasting service RTBF (14.3%), and the French Swiss broadcasting service SSR (14.3%).

TV-5 was inaugurated by the French Culture Minister, M. Lang, in February 1984. The programming is a structured 'best of' each of the member channels in French, and each channel is allocated one or two days per week to programme, according to their equity share. Total programme time averages 3 hours per evening. The programme emphasis of each channel is:

TF1   Entertainment, drama, adventure, current affairs
A2   Drama, adventure, entertainment
FR3   Entertainment, drama
SSR   Current affairs, music, arts, entertainment
RTB   Entertainment, current affairs, music/arts

In 1985, programme material was also supplied by the French broadcaster in Quebec, Canada.

TV-5 is targeted towards discerning viewers, with a cultural inclination in programme viewing. While TV-5 is in French, it is designed for viewers with a limited knowledge of the language, and is targeted European-wide. TV-5 is currently viewed on cable systems in West Germany, Finland, France, Switzerland, Belgium, Sweden, Norway, the United Kingdom, the Netherlands, Quebec (Canada), and North Africa, with an estimated European television household penetration of 2.5 million by December 1985.

During 1984, TV-5 received subsidies of $160 million from the French government while it was transmitted in the clear. Throughout 1985, it will carry no advertising and will be funded through contributions from the programme providers themselves. Sponsorship and advertising are both under consideration as alternative sources of funding.

3-SAT

3-SAT is a German rebroadcast service that is distributed via Eutelsat 1-F1, transponder 2 (east spot beam). It uses a SEL-PCM2 decoder. Equity in the service is shared among three German language national broadcasters: ZDF (71%), ORF (26%), and SRG (3%).

3-SAT offers experimental German language music and entertainment programmes for 3 to 6 hours per evening. The programmes are rebroadcasts of those shown previously on the individual channels and the channel is viewed on cable systems in Austria, Switzerland and West Germany.

Early cable viewer surveys indicated that there was little viewer support for this channel. It is targeted towards German language speakers, and there is a very limited subscriber base for such a service at present. It was the only satellite channel to be carried in the Dortmund cable experiment in West Germany in 1984, as it satisfied the North Rhine–Westphalian SPD Government's requirements of public broadcasting. 3-SAT is funded by 40 million DM from ZDF, 10 million AS from ORF and 1 million SF from SRG.

RAI

RAI is a relay channel which transmits via Eutelsat 1-F1, transponder 1, for experimental purposes. Its plans include a transfer to Olympus in 1987.

RAI intends to retransmit its current terrestrial programming from RAI 1 via satellite throughout Italy and Europe. RAI offers Italian general entertainment programming and news, broadcasting 7 days per week, both during the day and in the evening.

In 1985 RAI broadcasts via Eutelsat are used for experimental purposes only, and authorized reception is limited to one experimental station near Turin. Italy has no cable networks due to its proliferation of private television stations, and RAI has not resolved legal issues concerning copyright and retransmission which would allow it to be received elsewhere (the Italians have only paid for a domestic lease on Eutelsat 1-F1).

However, the RAI is currently negotiating reception rights elsewhere in Europe. The service began distribution on Belgian cable networks, which cover a significant population of Italians, in March 1985. The channel was reportedly popular in areas with large Italian communities who were re-transmitting the broadcasts illegally in early 1985 until the Belgian PTT took out an injunction to stop reception. The service was restored in July 1985 and is also seen on Swiss cable network. RAI has come under increasing competitive pressure from Silvio Berlusconi, who now owns three of the four major terrestrial private 'networks' that have been established. They provide popular entertainment programming, 80% of which is foreign, and the ratings of Berlusconi's Canale 5, Rete 4, and Italia 1 exceed those of RAI's second and third channels. As a result, the RAI must find new ways to compete for audiences and it views satellite-distributed services as a major possibility.

RAI via satellite is to be funded through a combination of advertising and financial support from the terrestrial RAI organization. Until it is officially acknowledged as a channel, however, the experiments will be entirely funded by the RAI.

Proposed services

The ITCA in the United Kingdom would like to launch a general English language entertainment channel as a pan-European channel to compete with Sky Channel and to exploit the market opportunities for distribution of programmes. The archives of the ITV companies together with the ITN news service provide the nucleus of an attractive service led by Central TV, a UK 'Super Channel' has a proposed start-up date of October, 1986. Unlike the other national broadcast services this channel is designed to pay its way and is a first indication of the way broadcasters can respond to the 'intrusion' into their markets of the new channels.

8.4.3 SPECIALIZED AND NARROWCAST SERVICES

As in the United States, cable offers the opportunity for programme providers to deliver specialized channels for select market groups. Music Box was the first channel to begin a European service followed by a religious channel, New World Channel, and a general public affairs channel, World Public News. By the end of 1985 one additional channel had been launched: Ted Turner's Cable Network News from the United States.

Music Box

Music Box is delivered throughout Europe via Eutelsat 1-F1, transponder 12, west beam, (non-eclipse protected), by Thorn EMI. Music Box has been established as a separate company, Music Channel Limited, with original equity partners being Thorn EMI with 50%, Virgin Records with 45%, and Yorkshire TV with 5%. The US based MTV is also seeking an equity share.

Music Box first began to broadcast programming on Sky Channel in February 1984 for several hours each day. It began broadcasting on its own channel on 12 July 1984 for 12 hours per day. The current schedules run 18 hours of programming comprising 6 hours of locally made product. The programmes are broadcast primarily in 1-hour segments which include Chart Attack, Gig Guide and Eurochart.

Of the programming 85% is composed of pop videos. About 14 videos are broadcast per hour and are shown on a priority basis: high rotation is 5 to 7 times per week, medium is 3 to 5 times per week, and low is 1 to 3 times per week. The other 15% of the programming is composed of video jockey links, advertising competitions, quizzes, music, news, and interviews. Programme production costs are about $1200 to $1700 per hour.

Music Box is targeted towards viewers of 15 to 25 years of age, primarily in continental Europe. It is currently broadcast in West Germany, the Netherlands, Finland, Sweden, Belgium, and the United Kingdom. Agreements have been reached with Denmark and Norway and negotiations are underway in Austria and France.

Music Box reached 2.2 million households at November 1985, with a country breakdown as follows:

Netherlands	853 000
Switzerland	328 000
United Kingdom	116 000
West Germany	67 000
Finland	102 000
Sweden	41 000
Belgium	700 000
	2 207 000

While it plans to increase to 7 million viewers by 1987, Music Box has run into a channel saturation problem, particularly in highly cabled areas, like the Netherlands, because many networks are full. Negotiations though included those with RTB in Belgium to use the airwaves in the day and late at night when RTB is off the air, with TDF for terrestrial relay in France and with the Danish PTT to be carried terrestrially in Denmark. The alternative to channel saturation for the new channels is to *create* paths of entry into the various markets.

Of the revenue for Music Box 80% is generated from advertising. The standard rate for one 30-second commercial spot is $420 ($1.20=£1.00), and blocks of 28 to 112 30-second commercial spots are priced at approximately $230 to $360 per spot. A 12-month booking reduces this rate to $175 per spot. Per hour 90 seconds of advertising are broadcast. This should eventually become a 4-minute block, with 2 minutes reserved for local insertions. Cable operators are also charged up to $0.30 per subscriber to supplement revenue. Music Box believes it will go approximately $11 to · $13 million into the red before operating profits are achieved. Operating costs for the first year alone could exceed $5.5 million. Transponder costs on Eutelstat 1-F1 are $2.4 million per year. With $165 000 necessary for downlinking in each country, the total costs per year are over $3.3 million. Another $1.1 million is necessary for video packaging, studio presentations, and purchase of concerts and specials.

New World Channel

The New World Channel is a Norwegian-based service that is distributed throughout Europe on Eutelsat 1-F1, transponder 4, the same transponder used by TV-5. The New World Channel broadcasts during the day when TV-5 is not using the transponder. It is operated by Hans Bratterud and his European Broadcasting Network which operates out of Oslo.

Transmission of the service began on 1 October 1984 for 2 hours of programming daily. It offers family viewing programmes, with a religious orientation, and is targeted toward Europe in general. In addition to Scandinavian programming, programmes are also broadcast in Russian and Eastern European languages. The service is advertiser supported, but has received most of its financing from religious organizations. By the end of 1985 service was received by 350 000 households in Norway, Finland, Switzerland, the Netherlands and the United Kingdom.

World Public News

Marc de Cock of the Independent TV Service in Belgium began the World Public News Service in Spring of 1985. It shares the Belgian transponder with ATN-Filmnet and intends to offer a full news, public information, news background service from 06.30 to 18.00. The news service is oriented towards a European-wide audience but does not contain any commercials. It is programmed with 13-minute information slots followed by 2 minutes of news headlines. The revenue is collected from the sale of the 13-minute programme time slots which is billed in ECUs, 715 ECUs per block for a frequency of three showings per week for a one-year contract. Rates are linked to the number of potential viewers of the programmes. The service began in Antwerp in July 1985 and by December was also received on cable systems in the Netherlands and Belgium.

Cable News Network

Cable News Network (CNN) in the United States, owned and operated by Turner Broadcasting Services, became Europe's first transatlantic video service. The service, transmitted via a transponder on Intelsat VA-F10, began in October 1985. It provides news programming to EBU news organizations with the intent of expanding its own full service in several years. The targeted 'market' for early entry is up-market hotel rooms. The service is offered to hotels for a price per room ($9 per month is a quoted price), being sold to the hotel owner in the same way as the morning newspaper. CNN is also expected to be a part of Thorn EMI's Galaxy Television package for cable operators, along with Thorn's other satellite services. The channel is targeted towards business people and the upper educated strata in Europe. It intends to supplement the subscription with revenue from advertising. Key markets are the SMATV and hotel markets of Europe.

ScreenSport

ScreenSport is an English sports channel distributed to cable systems by a half-transponder on Intelsat VA-F10, west beam, through a British Telecom lease. The equity of ScreenSport is as follows: W. H. Smith (19%); ESPN (20%); ABC (10%); RCA (10%); Ladbrokes (11%); CIN (10%); ICFC (10%); Robert Kennedy (5%); and others—John Wren, Richard Price, Richard Leworthy, and Chris Howland (5%).

Robert Kennedy founded ScreenSport in 1983. It battled with Cable Sports and Leisure to supply the first sports channel in a market that could only support one. ScreenSport emerged the winner and commenced transmission on 29 March 1984 for 6 hours each evening. It broadcasts from 17.00 to 23.00 Monday to Friday, and from 16.00 onwards over the weekend.

Most of the programming is pre-recorded, with ABC, ESPN, and RCA, through NBC, as the major suppliers. While ScreenSport has vowed to remain European in scope covering sports and events which other networks neglect, programming also covers American sports. Other programme suppliers include MacCormack's TransWorld International, with 4 hours per week for two years, and Cable Sports and Leisure Productions, the former cable channel consortium.

ScreenSport has covered several live events. On Saturdays, they broadcast a live programme with the results of all major sporting events and in 1985 began covering some UK football and tennis matches. ScreenSport has covered some events sponsored by Ladbrokes, an 11% equity shareholder in the channel. It plans to increase its coverage of live events and sponsorship as it is offered on more cable systems. In addition, it operates a sports teletext service from 14.00 to 17.00 daily.

In early 1985, ScreenSport began its path into Europe, first through Finland and then Sweden. A multi-lingual text system was also introduced to allow local cable operators to insert results and news in local languages as subtitling. In late 1985 it extended further into Europe offering a pay service in the Netherlands, using the same decoder as Filmnet/ATN. A multi-lingual text system was also introduced to allow local cable operators to insert results and news in local languages as subtitling. There is an opportunity for a sports channel in each country and ScreenSport hopes to use a combination of sponsorship and, primarily, subscriber revenue, even if collected through the cable provider, as its revenue sources. In the United Kingdom, where ScreenSport established its early marketing base, it has focused upon the rebuilt and new cable franchises, and is on offer on nearly every system. It has plans to expand its subscribership further throughout Western Europe.

ScreenSport is currently capitalized at $4.5 million. It has two primary sources of revenue: advertising and sponsorship. An average 30-second commercial spot costs $170, with special rates for prime-time and special positioning, but in 1985 there was no rush of advertisers. Options to improve the revenue position include selling ScreenSport as a 'pay' service to cable operators, i.e., to charge subscribers a small but not insignificant fee.

Other-planned and speculative, December 1985
Samuel Broadcasting had leased a British transponder on Eutelsat 1-F3 with the objective of beginning to deliver the *Wall Street Journal* daily market review as well as other programming from FNN. To add to this package the channel intends to introduce modular programme packages and to enter Europe through terrestrial links as well as the delivery to the cable systems. The launch failure of F3 has delayed this service.

There are also several speculative pan-European services which are under various stages of development. Visnews, the news service comprised of the BBC, Reuters, the Canadian, Australian, and New Zealand broadcasting corporations has plans to offer a 24-hour news service to the whole of Europe by satellite. The channel would operate 24 hours per day, and broadcast in English, French, and German. It is targeted towards the young, mobile professionals throughout Europe. Start-up costs are estimated at $2.3 million, and over $11 million for annual operating costs is expected.

Tony Hemmings plans to launch an English-language, pan-European business channel. The aim is to provide training programmes, industrial-sponsored films, corporate news, and information on trade fairs and other business information. Revenue is to come through advertising and subscriptions.

As trans-Atlantic satellite services, such as Orion and ISI, become legal, it is likely that more American programming will be distributed directly to Europe (Chapter 5). The American cable channels themselves may become full European services as well as with European partners.

8.5 National services
National services are targeted towards a specific country with a specific language group. They are distributed both by satellite and terrestrially. The critical start-up problem facing the programme providers is that, almost without exception, national markets are relatively small. This is primarily because the service is limited to cable subscribers or those paying for the programmes. An early tendency is to use the national market as a base for expanding into other European countries, usually on a region or even national market-by-market basis.

The Children's Channel

The Children's Channel (TCC) is distributed to the United Kingdom via Intelsat VA- F10, a transponder which it shares with Premiere, the pay-television movie service. The channel is owned and operated by Thorn EMI and Thames TV.

TCC began transmission on 1 September 1984 to cable systems in the United Kingdom. It is broadcast 8 hours daily, in 4-hour blocks, from 07.00 until 15.00, at which time Premiere begins transmission via the same transponder. The schedule comprises: US cartoons; adventure series; shorts; *Jack in the Box* magazine programme—nursery rhymes, fairy stories, animation, learning, and music; repeat of programmes from the day before. Slight variations are made during the weekend as older children are home from school. Programmes also change, depending upon term times and holidays.

Of the programmes 15% is original material, with an expected eventual increase to 30%. Thames TV is producing 40 to 50 hours for the first year, and Nickelodeon, the US children's cable service, provides 1.5 hours of programming per day.

The target audience varies at different times during the day. TCC is aimed primarily at pre-schoolers and their parents and children from ages 5 to 14. It is targeted towards cable households with children, although it is being offered with a total basic cable package.

TCC hoped to reach 2.5 million subscribers by the end of 1985 but the delays in cabling mean that this target will not be achieved. As of December 1985, the channel had about 120 000 subscribers, whereas 2 million subscribers are necessary for TCC to break even. Initially, TCC is transmitted only in the United Kingdom. However, it has long-term goals to broadcast throughout Europe.

Revenues for TCC are obtained through advertising, subscriptions, and operator fees, although advertising will be the mainstay. Advertisers have responded well to the channel because very few homes on television are devoted to children, and they have limited opportunities to reach this market.

TCC plans to broadcast 6 minutes of advertising per hour, but will begin at a one-minute average. Cable operators are charged about $0.30 per subscriber for the service.

Mirrorvision (formally TEN—The Movie Channel)

TEN—The Movie Channel (TEN-TMC) was a UK pay-television film service distributed via Intelsat VA-F10, west beam, through a British Telecom lease to United Cable Programmes. Equity in the original service was: Plessey (13.75%); Rediffusion (13.75%); Rank-Trident (13.75%); Visionhire (for sale) (13.75%); United Independent Programmers (Paramount, MGM, Universal) (45%). With Robert Maxwell's acquisition of Rediffusion, the sale by Plessey and Visionhire of their equity, and, finally, the liquidation of the channel in May 1985 the result was a 100% owned film channel for Maxwell under the banner of Mirrorvision Film Services Limited.

TEN-TMC began transmission on 29 March 1984, making it the first movie channel available in the United Kingdom. Initially, it broadcast programmes in two hour slots, from 16.00 until 02.00. Programming consisted of films and general entertainment, including soap operas, mini-series, or single special programmes.

Audience surveys indicated that TEN-TMC was considered more as a general entertainment channel rather than a film channel, which made it weaker competition for Premiere, the other UK-based film channel (see below). As a result, TEN reduced the amount of general entertainment programming broadcast, and rescheduled its films into 90-, 105-, and 120-minute slots. Its name was also changed to TEN—The Movie Channel, to better inform viewers as to its content.

TEN-TMC broadcast about 50 films per month—15 films new to television and 10 first-run cinema releases. Of the new films, three were for over 15 year olds, four for over 18 year olds, and three PGs. Each of these was broadcast five times per month. The UIP partners supplied 60% of the schedule, with the remaining 40% from the United Kingdom, Australia, and Europe.

TEN-TMC was available on Rediffusion's upgraded systems, which pass 1.6 million homes, as well as several other systems, and had about 40 000 subscribers in May 1985 (a fall from its peak of 50 000) when the American film companies withdrew their support. The revamped service, Mirrorvision, has changed the channel's focus from premiere movies only to include family entertainment and soap operas as well as adult movies from midnight to 03.00. Further, if Maxwell goes ahead on TDF, the service could have a European market focus (Chapter 5). In terms of the evolution of channels in Europe it is clear that economic realities and market demand factors are beginning to have some effect.

The tri-partite consortium UIP, which supplied programmes to TEN, has entered into an agreement in the Netherlands with publisher VNU to provide a pay-television channel called ATN. This has since merged with its arch competitor Filmnet (see below). UIP is pursuing agreements in other parts of Europe as well. They are in discussions with Orkla Industries in Norway, Sanomat Corporation in Finland, a Flemish publisher in Dutch-speaking Belgium, RTBF, and Les Editions Dupuis in French-speaking Belgium, and companies in Switzerland and Italy. In addition, UIP has discussed sales arrangements with Canal Plus. A Bertelsman/UIP pay-channel for German-speaking territories fell through in late 1984, leaving UIP out of a major publisher/pay-movie venture in West Germany. The production houses are reevaluating their position in Europe. Traditionally they have sold products to all buyers rather than be associated with a single channel.

Mirrorvision has maintained the same fee structure as TEN. The cost to cable operators is about $4.00 per month. They in turn are charging subscribers $9.00 per month to receive the channel. Advertising and sponsorship options may be necessary, given the low penetration and high cost structure.

Premiere

Premiere is a UK pay-television film service distributed via Intelsat VA-F10 west beam, channel 6, through a British Telecom lease to Thorn EMI Cable Programmes Ltd. Initial equity in the service is as follows: UK–Thorn EMI (41.2%); Goldcrest (9.8%); Columbia (9.8%); 20th Century Fox (9.8%); HBO (9.8%); Showtime/The Movie Channel (9.8%); Warner Brothers (9.8%).

Premiere was launched as the second film channel in the United Kingdom in July 1984, going live on to cable systems on 1 September. Each month 25 new films are broadcast, four times during the first week. The films are not shown again until 9 months later, to retain an air of freshness in programming, and they are shown about five times in total during a 10- to 12-month period.

The schedule runs from 15.00 until 03.00, beginning at 17.00 on the weekends. Programmes are in three segments: children's, family, and adult. Adult films are screened after 20.00, which contradicts the guidelines given by the Home Office. Many of the films are American, drawn from the American partners, as well as those provided by Thorn EMI and Goldcrest.

The initial market is about five upgraded cable systems, with a 135 000 home potential. Premiere began with 3000 subscribers. Surveys of households receiving Premiere and Mirrorvision indicate Premiere is the preferred service.

Premiere International, which excludes Goldcrest from the consortium, has entered into an agreement with Teleclub to provide films for its pay-TV service, and has plans to expand to France and Scandinavia, as well as finding a path of entry into the Dutch market. At present, the technical standards required for a terrestrial service and the lack of an available Dutch transponder leaves Premiere International out of the Netherlands, unless it becomes possible to downlink the UK Premiere channel with appropriate copyright clearance and language subtitles or dubbing.

While Premiere would like to offer one pan-European film channel, copyright limitations and non-simultaneous release patterns prevent using the same satellite feed. It may be possible to transmit the English-language service directly into some countries, notably, the Netherlands and Belgium, which would increase the economies of scale significantly.

Premiere is on offer in the United Kingdom for about $8.50 per month. Cable operators purchase the channel for $4.00 to $4.60 per subscriber per month. Premiere has offered a complete sales package to operators to encourage initial take-up of the channel.

Teleclub

Teleclub AG is a German-language film channel distributed via Eutelsat 1-F1, transponder 7, to some cable systems in Switzerland. A sister company, Teleclub Club GmbH has been formed to provide a similar service in West Germany.

The German film distribution company Beta Taurus is a shareholder in both groups. Equity in the Swiss Paysat group is as follows: Rediffusion AG (40%); SRG (15%); Beta Taurus (10%); Cable Vienna (10%); Metropolitan (10%); Telesystems (7.5%); Telsat (7.5%). Equity in the German consortium is: Beta Taurus, Axel Springer and Bertelsman (51%), and Premiere International (49%). The marketing distribution company, KMG, (Gesellschaft für die Vermarktung von Kabel und Satellitenprogrammen) is 33% owned by each of the German partners.

The operating company in Switzerland is callled Paysat. It comprises the four Swiss companies who applied for the Swiss Eutelsat F-1 transponder. Paysat began to use Eutelsat F-1 to broadcast to Zurich cable homes, under the old Teleclub label. It signed contracts with other Swiss cable operators through Teleclub, a terrestrial pay-television operation on Zurich cable (a cable system with 200 000 connected households). Teleclub was 51% owned by Rediffusion and 49% by Beta Taurus. The service never achieved significant penetration of households.

Programming for Teleclub is composed primarily of films similar to those provided by Paystin Switzerland. Film programmes are supplied by SRG. The channel is to be broadcast 24 hours per day. The monthly subscription fee is about $9.00 per month. Penetration had reached 20 000 by the end of 1985.

Filmnet/ATN

Filmnet/ATN is a film channel that began distribution in March 1985 via Eutelsat 1-F1 on the Belgian transponder. It was initially targeted to Dutch-speaking subscribers. Partners in the service include: VNU (United Dutch Publishing Company), Filmnet, Esselte, who has transmission rights to the Belgian transponder, UIP, and the United Dutch Film Company.

Esselte, a large video company based in Sweden, procured the use of the Belgian transponder on Eutelsat 1-F1 to provide a European film channel.

As Belgium initially did not authorize pay-television, Esselte first focused upon the Netherlands as its initial target. The channel received approval to transmit to Belgian cable networks in September 1985.

The film channel will broadcast an 8-hour film schedule, from 17.00 to 01.00, in addition to drama series, chat shows, and other entertainment. Filmnet also intends to broadcast a breakfast show, from 07.00 to 10.00. Filmnet, which is associated with the United Dutch Film Companies, and Esselte Video will provide most of the programming.

The Netherlands service costs about $8.00 to $10.00 per month and covers an area with a potential of 4.6 million cabled households. Esselte estimates that approximately $11 million are necessary to start up the operation.

Esselte has long-term plans to provide a multi-lingual channel, focusing upon Scandinavia and the rest of Europe. It has developed EMSS—Esselte multi-subtitling system—to provide subtitles for the films in French, Flemish, Swedish, Danish, Finnish, and Norwegian, and already has pay-television programmes in Finland and in Denmark.

After Belgium, the channel is targeting Sweden and Denmark, through the Danish pay channel, and an agreement has been reached with the Norwegian company Nissen Lie Consult and Esselte/Filmnet to form Central Film, a joint company, to pursue possibilities for the service in Norway.

W. H. Smith ventures

The most extensive role in narrowcasting has been taken by the UK retailer W. H. Smith. The stated strategy is to be involved in the entertainment side as well as the interactive services side of cable and satellite. Their stated intention is to find proposed services by a combination of advertisements, sponsors, and cable operators' subscriptions, acknowledging that targeted services on cable cannot hope to attract advertisers' dollars in the way that broadcast networks can.

In addition to equity in ScreenSport (see above), W. H. Smith plans to be involved in four cable channel concepts.

Lifestyle In a venture with Reed International (25%), Yorkshire TV (20%), and TVS (Black Rod) (20%), in which W. H. Smith has a 35% equity stake, W. H. Smith proposes to launch a daytime satellite channel on Intelsat VA-F10 from November 1985. Lifestyle share a transponder with ScreenSport. Fashion, beauty, interviews, healthcare, leisure- and family-oriented topics will be covered. The target group will be women and the morning programmes will be informative and entertaining (e.g., soap operas). A price of $0.50 per basic cable subscriber per month is being asked from the cable operators wishing to take this channel. Initially, the United Kingdom is the chosen market but the wide appeal of the channel and the need for a large audience base will require W. H. Smith to contemplate the European market for some or all of the programming; Sweden, the Netherlands and Scandinavia have been targeted. As with other ventures, equity participation is changing—British Cable Services and D. C. Thompson have replaced Reed International.

Arts Channel   The British Cable Programmes propose to offer an Arts Channel for subscribers in the United Kingdom who want serious programmes—documentaries, drama, and music. In March 1985, W. H. Smith took a 30% equity stake in the Arts Channel, which is buying in and producing its programming. The locally produced material currently costs about $3000 per screen hour (including repeats), indicative of a price of $10 000 per hour, the channel's stated objective. The market is that group of people not satisfied by BBC2 or Channel 4. Other investors include TVS and Commercial Union. The proposal is to share the transponder with ScreenSport and Lifestyle in 1986.

Videoline   W. H. Smith proposes to use the television medium to launch a shopping channel. Initially, the channel will screen 'informercials'—long versions of advertisements, usually with significant information content.

Telesoftware channel   W. H. Smith plans a national service to deliver computer games, software, and data into the home by cable.

Lifestyle   A second Lifestyle service has been planned in the United Kingdom by a group led by Patricia Williams and Sue Francis, former editors of *Broadcast* magazine. Both Lifestyle groups need the Rediffusion network in the United Kingdom to get any penetration at all in the short term. The decision in July 1985 for Rediffusion to carry the W. H. Smith Lifestyle channel put Lifestyle back into the market-place seeking financial backers and programming partners.

Satellite TV

Satellite TV (STV) has taken a lease on Telecom-1 to provide an entertainment channel to cable operators and terrestrial broadcasters. It is proposed to start in early 1986, is France's first private broadcast company, although funded by the state agency Mission Cable, and has targeted the 24-45-year-olds. Cable operators will be charged $0.60 per connected household. National, regional and local advertising will be included.

Other

A number of other channels have been introduced into Europe. A German-speaking music video channel, Musicbox, was allocated a transponder on Intelsat V by the Bundespost. Programmed by KMP (Kabel Media Programmgesellschaft) the service began as an advertising supported channel on the Berlin cable network with other German networks to follow.

A new movie channel, Home Video Channel, is being offered in the United Kingdom. Unlike existing movie channels, it is a videocassette-based channel for distribution as part of the basic cable tier. Atlas Leisure Corporation, which bicycles videos to hotels in the United Kingdom is proposing to deliver its entertainment programmes, much of it video/music, by satellite. Atlas also manufactures TVRO equipment and is targeting the SMATV market.

Le Monde, Agence France-Presse and Gamma-TV are developing a satellite delivered French TV news.

8.5.2 TERRESTRIAL SERVICES

There are several programming services which are distributed to households directly through the use of the broadcast airwaves, aerials, and decoders. The first of these services is in the French language and there may soon be a proliferation of such channels if the French government implements the Bredin report.

Canal Plus

Canal Plus is a terrestrially distributed film service for France. Its equity is as follows: Havas (42%); Consortium of national banks, headed by Société Générale (20%); Compagnie Générale des Eaux (10 to 15%); L'Oreal (10%); Daily Newspapers–Havas clients (5 to 10%); 10% as yet unallocated.

Canal Plus began transmission in November 1984. Eventually, 20 hours of programming will be broadcast from 06.30 to 15.00 Monday to Thursday and for 24 hours on Friday, Saturday, and Sunday. Programming is centred around films, and includes sport, dramatic series, special events, news, and breakfast television.

Canal Plus has restricted quotas on its foreign films; 60% must originate from the Common Market and 50% must be from France. Other programming types have no quotas. Canal Plus has also entered into several joint ventures to produce their own productions.

To fill the annual schedule 320 programmes will be necessary. Films will be broadcast two to three times in 2 weeks, and unscrambled trailers will be broadcast for 3 to 4 hours daily. The signals are scrambled and broadcast on both 20 VHF and 37 UHF transmitters.

Initially, Canal Plus began transmission in the Paris and Rhône–Alps areas. It is hoped to cover half of France by the end of 1985, and 90% of France by the end of 1987, reaching 1 million subscribers by the end of 1986 and 1.5 million by the end of 1987. Eventually, it is hoped to broadcast via satellite, covering all of France. Breakeven will be reached with 700 000 subscribers.

Equity for Canal Plus is approximately $14 million. Financing is both public and private. It is hoped that 5% to 10% of the revenues is gained through programme sponsorship, while the remaining revenue will come from subscriptions. Half of the revenue will be allocated to running costs; 25% will be allocated to film rights purchase or programmes supporting the French production industry; 25% will be allocated to other programme sources.

Subscription fees are about \$12.00 per month, for 6-month or 1-year subscription blocks. \$58 is needed for VHF aerials in those areas where transmitters will use this frequency. A \$44 refundable deposit is required for the decoder. In addition, collective aerials can also be purchased for a block of flats for about \$263. In addition, the 6.6 million television sets made prior to 1981 lack an input jack for the decoder and an \$84 adaptor is necessary.

The channel began auspiciously, averaging 3000 new subscribers per day, but a series of malfunctioning decoders and the President's decision in early 1985 to permit local private television halted this rapid penetration.

An agreement with BLIC (Bureau de Liaison des Industries Cinématographiques), the French cinema organization, allows Canal Plus to broadcast films on Wednesday, Friday, and Sunday evenings in prime time, a first for French television.

Penetration picked up in mid-1985 and by the end of 1985 there were 650 000 subscribers—on target to break-even.

Telecine

Telecine is a terrestrial film channel of the operating group Telecine Romandie, a joint venture among SSR, Swiss Film and Video Production Companies, Cambridge Film Group (US), and Cinevision (Belgium). It began operations in December 1985, transmitting to cable systems from a broadcasting tower located at Mount La Dole on the French–Swiss border. The signal will then be relayed by microwave to the cable systems. The channel offers film and entertainment in French by three pay categories:

Category 1   Films from 17.45 until 23.30
Category 2   Adult entertainment from 23.30 until 24.00 during the week and 23.30 until 03.00 on the weekends
Category 3   Sports programmes on Saturday and Sunday evenings and two weekday evenings, including English and Brazilian football, American basketball, and ice hockey; music for one hour daily; children's programmes for 4 hours on Wednesday and the following Tuesday, plus 4 hours on Saturday and Sunday mornings; 2 hours per week of original programming

Over 40 films per month, including 15 first-run showings on Swiss television, will be broadcast. Of the programming 40% will come from France, some from Canal Plus.

Telecine Romandie is targeted towards the French-speaking audience in Switzerland, a potential 500 000 viewers. Revenue for the service will be generated through subscription fees. Each category of programming will be available for $2.00 to $10.00 per month. The decoder necessary for descrambling will be rented for about $5.80 per month. Telecine Romandie will need $7.2 million to begin operation.

8.6 Local programming services
With the advent of new cable systems, cable services distributed locally from the head end have also begun to emerge. These channels are usually targeted towards a narrowly defined audience which eliminates the need to be distributed via satellite. Several channels have been in operation for a number of years and have provided examples on which the new services can be based.

The first large European cable system to offer a local service was Helsinki Televisio (HTV) in Helsinki, Finland, established in April 1973. A second major local system, Teleclub, was established in Zurich, Switzerland, by Rediffusion, Switzerland's largest cable system. A similar pay-television service, VideoCinema, was developed by the Hachette group in France for a privately owned estate in Paris.

New local channels are being developed for the cable systems which have recently been constructed. Most of these channels are targeted towards specific audiences and are delivered by videocassette and computer directly from the head end.

Much of the new development of local services has been initiated in the United Kingdom and West Germany because they are the furthest along in the construction of new systems. In the United Kingdom, channels providing computer games and software have been proposed by British Telecom, W.H. Smith, and Thorn EMI. Computers at the head end would distribute a monthly service to households. Cultural, educational, and women's channels have all been proposed, in addition to some interactive channels for banking and shopping.

In West Germany, recent reversals in media policy have paved the way for private programming on cable networks to be permitted in most Lander. In addition to private satellite-distributed services, the new cable systems will now be able to offer local commercial cable services. Bidders for the spare channels have already emerged, with ideas for news, magazine, and film channels.

The French decision to allow local and regional television (Chapter 2) has attracted a number of large corporations jockeying for positions in the prime television markets, particularly in Paris, with an aim to become a programme provider for the local stations.

Major groups such as publishers Hachette, Publicis, Les Editions Mondiale, Pathé-Cinema (a film distributor), as well as the broadcasters RTL and Télé-Monte Carlo/Europe 1, announced their intentions to bid for the national networks and there were hundreds of requests for the individual local stations. Only one network has been allocated, to Berlusconi, an Italian TV magnate and a consortium of French industrialists. STV has taken a transponder on Telecom 1 and begun serving programmes such as Music Box to create a cable channel. The CLT group, the early front runner, had not resolved its position with France by the end of 1985.

Hersant, owner of *Le Figaro*, created Teleurop (TVE) in January 1985, a channel which is news and entertainment oriented. It proposes to broadcast 18 hours per day, beginning in 1986, be advertiser supported, and offer 4 hours per day to local stations to comply with the government's requirements for a national network.[4]

8.7 DBS channels

DBS programming services are also under contemplation by prospective suppliers. As yet, the majority of the proposed services are from the operators of the DBS satellites themselves. Because many of the hardware aspects of the DBS systems are still vague, and it is uncertain as to how many satellites will actually be launched, their future programming is somewhat indefinite. However, DBS operators do have ideas about what services they would like to have broadcast via their satellites.

TDF-1, the French DBS satellite, will have four DBS channels. Three have been allocated to programmers—Berlusconi-Seydoux, Robert Maxwell, and the French cultural channel. Berlusconi proposes to transmit entertainment programmes in French, Italian, and English, with Maxwell transmitting in English. TDF is taking on a European focus.

The West German DBS project, TV-SAT, is beginning to solicit candidates for its four channels. The proposed configuration at the end of 1985 was ARD Eins-Plus, a new German-language culture programme, 3-SAT, currently on Eutelsat, and two private programmes including SAT-1. No decision is expected until the Länder agree their media policy.

In West Germany, the state broadcasters propose to offer regional channels and the publishing group Westdeutsche Allgemeine Zeitung is looking for a path of entry to offer a general entertainment channel.

In Benelux, Dieter Minning, who had the lease on the Dutch transponder on Eutelsat 1-F1 but was forced to give it up when his finance fell through, plans to provide an entertainment channel for the Dutch market with possible extensions to Europe.

8.8 Conclusion

The pioneer channels—those running by the end of 1984—have emerged in response to the perceived market demands. In the United Kingdom, reputed to have the most entertainment-oriented television in Europe, the new channels aimed at that market are special-interest channels. Shake-outs have already occurred, new consortia are forming, and certain trends are becoming evident. The first phase of new television media development is summarized in Table 8.1.

Table 8.1 New television media development

Type of channel	Channel name	Language/language market	Estimated subscriptions December 1985
Entertainment	Sky Channel	English/European	5.0 million
	RTL-Plus	German	900 000
	Europa	Dutch/ Multi-lingual	
	SAT-1	German	500 000
Rebroadcast	TV-5	French/European	2.7 million
	3-SAT	German/German territories	350 000
	RAI (relay)	Italian (experimental)	
Special interest	Music Box	English/European	2.5 million
	New World Channel	Scandinavian	
	ScreenSport	English/UK	120 000
	The Children's Channel	English/UK	120 000
	Mirrorvision Channel	English/UK	50 000
	Premiere	English/UK	20 000
	Teleclub	German/Swiss	20 000
	Filmnet/ATN	Dutch	
Terrestrial	Canal Plus	French/France	650 000
	Telecine Romandie	French/Swiss	

Trends

Some implications are already clear.

Cable and DBS will provide additional windows in the existing patterns for rights releases   Using the USA as a model, the basic release pattern for a movie from a major studio is shown in Table 8.2.

In Europe, the new windows will be:

1. National VCR local language rights.
2. Language market advertiser-supported cable rights. The extensive use of communication satellites for programme distribution makes nonsense of individual country market rights.

Table 8.2 Basic release pattern for a movie

Outlet	Release time (following theatre release)
US theatres	0
Foreign cinemas	+ 6 months
Worldwide video cassettes	+ 4 to 10 months
Pay-television—US	+ 12 months
Network television—US	+ 24 months
Overseas networks	+ 30 months
Off-network pay-television	+ 50 to 60 months
Syndication	+ 60 months

3. National (or possibly language) market rights for pay cable.
4. National, language, and European-wide pay-per-view rights.
5. National and language DBS rights for programmes.

Cable and DBS Channels will create new programming opportunities The increased number of channels will eventually allow narrowcasting, programming aimed at small well-identified target groups, to be economic. In the interim, there will be made-for-television movies, local or regional programming, concert, theatrical, and other live specials and spectaculars made primarily for television, creative video-music clips and shows, coverage of additional national and international sporting events, and increased demand for entertainment—soap, drama, and mini-series—as well as children's and music programming. For example, News International, the majority equity shareholder in Sky Channel, signed an agreement with Albert Frère's Banque Bruxelles Lambert of Belgium to establish a joint development company.

The economic constraints necessary for non-public broadcast channels to become established is already producing a lean and inexpensive approach to production. Per hourly production costs far below broadcast television costs are the norm for Sky Channel, Music Box, and The Children's Channel. Technological developments (such as $\frac{1}{2}$-inch Betacam machines) will help create the pool of output which may eventually force down the price of production by the broadcast entities.

Advertising would seem to offer the best options for funding these channels but the fragmented audiences, within and across nations, suggests that a 'pooled' or 'packaged' approach will be necessary for advertising campaigns.

Cable and DBS channels will eventually challenge the dominance of established broadcast media National broadcasters have already begun positioning themselves for these opportunities. The proposed UK super channel includes the ITV companies; ZDF (West Germany), ORF (Austria) and

SRG (Switzerland) are offering a cable relay channel; TV-5 is an amalgam of programmes from the French, French–Swiss, and French–Belgian networks and Europa TV is an arrangement including Dutch, Irish, Italian, Portuguese and West German (ARD) broadcasters.

In the United States, the three networks have responded to the cable challenge by further strengthening their programming with mini-series and made-for-television movies. In Europe, the simultaneous introduction of DBS and new cable systems will erode the traditional network audience base more rapidly than in the United States. There will also be increased transmission hours and programming for 24 hours per day—another new phenomenon.

Although national–cultural differences, as well as regulatory barriers, will prevent homogenized American entertainment and film packages from being the only new products, the economics of film and series production will stimulate a commercial approach to programming. The first trend is in co-productions, and that is under way already among European countries and between Europe and the rest of the world. Second, there are the technological advancements that will allow simultaneous transmissions of multiple sound tracks. Already dubbing or subtitles, often through teletext, are common for film products. Third, there will be the realization that satellite—either DBS or communications—offers a path of entry to European and eventually world markets, rather than just national markets. This exists for specific events such as the Olympics. This opportunity will itself create a demand for certain products. Already there is talk of a world news channel.

Numerous ventures and joint ventures were announced in 1984 and began in 1985. Further channels are planned with the introduction of the various DBS satellites.

Satellite-delivered cable services offer the potential of a pan-European service The introduction of satellites has allowed pan-European channels to develop that might not otherwise have large enough audiences to sustain them simply on a national level. This is particularly true for specialized services, and it is likely that more of these services will develop as they did in the United States.

The undoubted success of Sky Channel depicts the vacuum that exists in European broadcasting—a dearth of entertainment-oriented programming and a lack of programming directed at the young. The advertisers' support, coming to the channel in 1985, although slow to develop in 1984, is a further indication of a market gap—this time for television advertising of the myriad of pan-European products. Casualty rates will be high—another US phenomenon.

Possible moves from Eutelsat to DBS satellites Unlike the United States, the KU-band satellites began television distribution prior to the C-band satellites. The practical distinction between KU-band and DBS satellites is not always clear, and it is difficult for companies to develop both satellite hardware and software. As a result, it is possible that the services that are currently delivered via a KU-band satellite such as Eutelsat 1-F1 will be the same services delivered on the DBS satellites. Every existing channel on Eutelsat has been reported to have considered some satellite transponder relocation option—be it to the medium-powered satellite or to a DBS satellite.

We have shown that, even if all DBS projects do not succeed, transponder availability will not be the key problem. What will become apparent in Europe over the next three years is that *satellite positioning* is critical.

Emergence of packaging The difficulties of selling to small markets and the prospect of selling to even smaller ones, namely SMATV, where the equipment as well as the programmes needs to be sold, has encouraged product packaging. A programme including ScreenSport, Sky Channel, Music Box, The Children's Channel, and Mirrorvision (formerly The Movie Channel) was sold to the Rediffusion cable networks in 1984. The introduction of legalized SMATV in the United Kingdom led to a Thorn EMI group, Galaxy, selling Premiere/The Children's Channel and ScreenSport with Music Box if an additional antenna is installed. Mirrorvision and Sky Channel are offering their programmes as single-channel options.

In West Germany, a similar concept developed by Fuba has a more interesting conglomerate of programmes. Eutelsat 1-F1 carries SAT-1, RTL-Plus, and Teleclub in the German language as well as Music Box and the English-language entertainment channel Sky Channel. This provides the core of a potentially very popular package for the German-speaking territories.

Big corporations are moving in The publishers across Europe see television as a natural extension of their businesses. With cross-ownership rules being applied mainly at the local level there is scope for the publishers to take positions in the international channels. Positioning ultimately will be critical and most publishers are staking an interest in a channel (the software) rather than the cable infrastructure or the satellite systems.

The market is evolving 1985 saw the first casualty in Europe as The Movie Channel went into liquidation, although it was restored quickly with a new owner. Further, the equity position in many channels is extremely fluid. New partners are entering and old ones are rethinking their roles. And the industry is only three years old! Additionally:

1. Pay-channels require a large socio-economically desirable population. With fragmented cable systems, small numbers of well-developed advertiser-supported channels to propel cable penetration and high VCR penetration, and programming rights sold on a national basis, the take-up of pay-channels has been very slow.
2. Narrowcasting, too, needs a large population base from which a relatively small percentage take-up results in a viable service. So far only Holland and Belgium even offer that possibility.
3. Television advertising in Europe has not been developed to capacity (from a demand perspective) on the broadcast networks, and as the new channels get 'sellable levels' of audiences, the advertisers will come.
4. Pan-European products are becoming more prevalent and the worldwide trend is to regional and global marketing opportunities. Corporations are looking for ways to maximize advertising return and the pan-European sell becomes possible once penetration across a sizeable number of countries approaches 2 to 3 million households.
5. The structure of many of the channels is still evolving. The Movie Channel is the most extreme example, but Rediffusion, Teleclub, SAT-1 and ATN/Filmnet have all reorganized. Groups will begin to form ventures across national boundaries for programming (e.g., RAI, Antenna 2, and Channel 4) and for efficient selling of advertising.

Narrowcasters beware in the short term Narrowcasters need to plan their entry to this market carefully. Sky Channel in the United Kingdom provides viewers with a core of entertainment programmes around which narrowcast programme concepts have been built—sport, music, and children's programming. Not only does this mean that the channel caters for all tastes, and hence is eligible for all sources of available advertising dollars, but until the market matures Sky Channel, by virtue of its sheer penetration, will be a better venue for specialized programming than many of the narrowcasters.

References
1. Simon Baker, 'Finnish connections', *Cable and Satellite Europe*, October 1985, p. 20.
2. Commission of the European Communities, *Interim Report: Realities and Tendencies in European Television: Perspectives and Opinions*, Commission of the European Communities, Brussels, 1983, p. 71.
3. The following information on European software services was gathered from a variety of sources, including: *Advertising Age Magazine, Broadcast Magazine, Cable and Satellite Europe, Cable and Satellite News, CableVision, Focus Magazine, Interspace, Marketing Magazine, New Media Markets, Neue Medien, Ogilvy and Mather Euromedia, Satellite and Cable Television News, Satellite Communications, Satellite News, Satellite Week, Television Weekly, Variety Magazine*; and advertising rate cards, annual reports, conference proceedings, informal conversations, press releases, publicity brochures.

9

Information services

9.1 Introduction

The primary focus of this book so far has been on new media entertainment services in the European household that are accessed via a television set. The television set has also been instrumental in bringing information services into the home: teletext, videotex, and personal computers.

Numerous terms have been introduced to describe the one-way and two-way electronic transmission of data and information services to the household. The CCITT (International Telegraph and Telephone Consultative Committee) has used *videotex* as the generic name to describe the provision of two-way information services and *teletext* for one-way services.

Teletext is broadcast text which is inserted into the vertical blanking interval of the television picture (or could take up the whole frame if there was no picture transmitted, as in the case of a fully dedicated text channel). It is a one-way continuous flow of information from the broadcast station to the household. The supporting text is cycled by the broadcast station, and viewers 'tap' the television signal at any time during its transmission, i.e., a limited amount of material must continually be retransmitted. The text may be updated frequently and a variety of games, computer software, advertisements, and even system messages can be delivered. In practical terms, teletext is a simple enhancement to television. It does not require additional stand-alone hardware and if the teletext chip set is integrated within the television set, teletext adds only a marginal contribution to the total price of the receiver.

Videotex, on the other hand, involves a two-way flow of information. This requires the user to have not only a device for receiving the signal, usually the television set, but a communication line out of the household, which can be either the telephone or, when installed, two-way cable systems. As telephones in Europe adopt time-sensitive pricing, even for local calls, the system has an ongoing usage price attached to it. Sophisticated autodial modems with memories, which can enter prearranged videotex systems in off-peak times, and videotex terminals with large electronic storage capacity are all being constructed to reduce the operational cost, but these require an educated consumer. As a result, videotex penetration will be slower than teletext.

The personal computer (PC), however, has provided the path of entry to the household for a variety of programming and information services. Initially, it was a stand-alone device used for home computing by hobbyists and enthusiasts. The introduction of national packet switched networks and, in the United Kingdom, pages of teletext devoted to computer programmes which can be downloaded directly to the home PC, is increasing the networking capabilities and is making the distinction between videotex and personal computers somewhat academic.

The subject of personal computers is not addressed extensively in this book as the primary focus is broadcast media and broadcast-related services. Of course, the merging of computing and communication technologies will open up the new possibilities in interactive data and video services in the household.

9.2 Teletext and videotex development

Teletext and videotex are technologies which originated in Europe. The first full commercial videotex service, known as Prestel, was initiated by British Telecom (then the British Post Office) in 1979. British Telecom, as the driving force in the establishment of the technology, took on the role of common carrier of the service. Teletext services emerged at around the same time, again in the United Kingdom.[1] They were first introduced by the British Broadcasting Corporation (BBC) and, separately, the Independent Broadcasting Authority (IBA). Nationwide public teletext services began in 1976 under the names Ceefax (BBC) and Oracle (IBA).

Simultaneously, a parallel development of the two technologies was taking place in France. The CCETT (Centre for the Study of Telecommunications and Television), a research centre to develop communication technologies jointly for the French PTT and Télédiffusion de France (TDF), was responsible for Antiope, a system incorporating both broadcast teletext services and interactive videotex services.[2]

The British and French systems emerged under different standards and were not compatible. The key differences in the two teletext systems were in the methods of signal transmission. Synchronous or fixed-format signal transmission, where there is a fixed relationship between the transmitted data and the corresponding characters on the screen, was adopted by the British and an asynchronous transmission system was developed in France. Both systems use a packet structure for digital information.

Further, the method used to transmit display characters—colour characters —was also different. The British used a method of serial picture attributes where the information and attribute code are transmitted sequentially within the data stream and the attribute code is then displayed as a space in the background. The French used picture parallel attributes, which avoids this problem by transmitting both the picture and its code together.

By 1981, numerous competing standards had emerged from national systems in Japan, Canada, and the United States, as well as the United Kingdom and France. The 1981 European Conference of Post and Telecommunications (CEPT) adopted a single European videotex standard. The standard, which has been ratified by 26 countries, defines a common alphamosaic system which interprets both the Prestel and Antiope systems and incorporates both serial and parallel attributes in the standard. Its implementation, however, is proving to be more difficult than anticipated.

9.3 Teletext services

Teletext technical trials and services are now under way in most European countries. By late 1985, around 4.5 million television sets with teletext were in place in European households. In the United Kingdom alone, the market is growing at a rate of about 60 000 sets per month. The British-developed standard, now referred to as World System, is used widely and has become the *de facto* European standard. The French system, Antiope, is the national system for France and it is also being used in conjunction with Ceefax in Belgium and Switzerland (Table 9.1).

In all European countries, teletext services or trials are being conducted by national broadcasters. Most services are financed out of special government grants, existing broadcast licence fees, or annual government allocations to broadcast stations. In France, however, Télédiffusion de France (TDF) is charging information providers, thus creating a source of revenue, and is also offering closed user group 'pay' services. Independent Television (ITV) in the United Kingdom is collecting revenue from the sale of pages for advertising and has introduced closed user group or restricted access pay services over the broadcast airwaves. Air Call, a mobile radio company, is using Oracle in 1986 to deliver subscription teletext services (financial information and data for betting shops). The BBC also proposes commercial applications by leasing Ceefax circuits to private organizations.

9.4 Growth of teletext

Teletext is not viewed as an essential consumer item, but rather as an add-on to colour television. Consequently, growth follows the replacement television market; it is not of itself the primary reason for television purchase.

In the United Kingdom, for example, sales and leases of teletext sets accounted for about 10% of the total shipment of television sets to dealers

Country	Name of service	Generic technology	Date service commenced	Service provider	Financed by	Magazine size (pages)	Number of terminals (end 1984)	Availability
Austria	ORF-Teletext	World System (WS)	1980	ORF	ORF (government)	200	200 000	08.30 to end of transmission
Netherlands	Teletekst	WS	1980	NOS	Government (as an experiment)	100 (+)	300 000	08.30 to 12.00
Sweden	Text-TV	WS	1980	SVT	Sveriges TV (public corporation) and existing licence fee	120	350 000	During transmission; 10.00 to close
United Kingdom	Ceefax Oracle	WS	1976	BBC ITV	Licence fee Ad revenues (ITV)	200 200(+)	2 600 000	06.00 to close
West Germany	Videotext	WS	1980	ARD/ZDF	ARD/ZDF revenues	200 (+)	300 000 (+)	16.00 to close
France	Antiope	Antiope	1977	TDF	Fees from information providers and subscription fees	2000 (+)	100 000	Varies among stations; mid-morning to close
Belgium	Teletekst Perceval	Antiope	not available	BRT RTBF	BRT revenue RTBF revenue		150 000 2 000	
Norway	Text-TV	WS	Trial only 1979	NRK	Licence fee	300 (est)	80 000	During transmission times
Finland	Text-TV	WS	1981	Oy Ylesradio (YLE)	Finnish radio	100 (est)	20 000	
Denmark	Tekst-TV	WS	Technical tests only	DR	DR licence fees	100 (est)	100 000	
Switzerland	Teletexte	WS	1981	SSR	SSR/government	100 (est)	200 000	
Italy	Televideo	Antiope	1985	RAI	RAI		10 000	Beginning in 1985
Spain	Teletext	Antiope	1985	RTVE (Radio Television Española)	Subscription	—	—	

* From communication with National Broadcasters-Videotex International[3] and ITVN.[4]

in 1981, a figure which rose to around 30% by 1984. In Austria, teletext-installed television sets account for 14% of all television sales and in the Netherlands, West Germany, and Switzerland, the figure is around 20% of television sets.

If we view the replacement (and growth) markets for colour television sets as about 10% per annum of the installed base, and if teletext sets account for about 30% of those sets, it will take 20 years to achieve market saturation (Table 9.2).

Table 9.2 Estimated teletext penetration by percentage[1]

Country	1984	1987	1990
Austria	3	11	24
Belgium	4	10	14
Denmark	3	9	15
Finland	3	6	14
France	1	10	25
Netherlands	4	8	15
Norway	3	10	18
Sweden	9	16	30
Switzerland	4	8	12
United Kingdom	8	25	55
West Germany	2	10	25

A further stimulus for penetration is the price of a television receiver with teletext. The price of the world standard teletext chip set has dropped significantly over the past 3 years so that the cost of the chip set is now only around $2. The cost of a teletext television receiver (either purchased or rented) has reflected this and has declined substantially in comparison with a non-teletext receiver. This reduction has been true for all countries where teletext is integrated within the television set.

9.5 Magazine size and content

It is difficult to be precise about the size of a teletext database. Television stations are allocating more vertical blanking internal lines to carry the services (now averaging four to six VBI lines across Europe), frames are continually being updated, new material is slotted in at various stages, and a 'page' may involve a number of 'frames' of data or subpages.

In most cases, teletext databases are divided into magazines of 10 to 30 pages. The general content for systems in Austria, Belgium, the Nether-lands, the United Kingdom, and West Germany is compared in Table 9.3.

Table 9.3 Database content of six teletext systems*

Austria	Five magazines
	Service information (stock market, weather, airport, traffic, metal prices)
	Subtitling (for deaf and general public; for example, operas)
	Radio and television programmes (today's and tomorrow's)
	Games (chess, riddles, horoscopes)
	News, headlines, data line to the BBC, plus many subpages such as snow forecasts and other details
Belgium	Eight magazines
	News, sport
	Weather, traffic
	Television–radio
	Press preview
	Finance jobs
	Leisure time
	Educational and consumer information
	Subtitles
Netherlands	Eight magazines
	News, weather, environmental warnings, subtitling, daily bulletins for children
	Consumer information such as purchasing, gardening, plants, foodstuffs, animal licences and vaccinations, general news headlines
	Financial information, inland shipping, spot market, foreign currency, gilt-edged shares, news flash, alarm information emergency (calls to individuals)
	Time-out (museum, films, theatres, concerts, puzzles, hit parade, page for viewer letters)
	Weather (long term—up to 5 days) and traffic in Holland and Europe, tidal information, railways, airport arrivals
	Sports news
	Television/radio programmes guides
	Broadcasting societies in Holland
Switzerland	Around 150 pages including:
	National and international news
	Stock exchange
	Sport
	Television and radio
	Theatre, hotel, restaurants
	Airport information
	Calendars, letters, hit parades, etc.
	General, consumer, educational
United Kingdom	Compiled from BBC1, BBC2, and ITV:
	Alphabetical indexes
	News, news headlines, and news flash
	Finance/business
	Sports
	Television programme guides
	Review/entertainment/art

Table 9.3—continued

	Regional entertainment
	Weather/travel news and road news
	Fun/puzzles/leisure
	Children's pages
	Food/farm/garden
	Consumer news
	For the deaf
	Alarm page
	Advertising (ITV only)
	What's on in your area
	Telesoftware
West Germany	News and news flash
	Today's parliament
	Television/radio programmes for the various networks
	Money market
	Consumer advice and cultural information
	Traffic and weather: local, national, Europe
	Previews
	Advice to parents on television programmes
	Radio stations in West Germany
	Games of chance (Lotto), quizzes
	Sports news
	Press previews of the next day's newspapers

* Published teletext database content of various broadcasters.

Sweden and Denmark developed their systems primarily for the hard-of-hearing and also have extensive subtitling services. The other content seems to be a variation of what is available in the United Kingdom, with the notable addition of a large amount of 'consumer-oriented' information from the respective Departments of Consumer Affairs.

The evolution of the various teletext magazines show remarkable similarity. A key ingredient is radio news-type information which is rapidly changing and topical.

Most systems are moving from national (only) magazines to regional magazines with local content and, in the case of Oracle, local advertising. They also include alphabetical indexes.

Broadcasters are developing novel ways to increase the volume of information and increase the interest in the system. They include:

1. Increasing the number of VBI lines.
2. Arranging different cycle times for high- and low-priority pages.
3. Including certain pages (or even whole magazines) at certain hours.
4. Increasing the number of subpages. Often up to 20 subpages can be linked to each page of a 100-page teletext database. This is useful, for example, on Saturday evening in the United Kingdom, when the teletext service carries the results of all English and Scottish soccer league games.

Manufacturers are now producing teletext receivers that can be programmed to hold more than a single page so that waiting time for the cycled page to appear will decrease. In addition, the BBC is well advanced in its telesoftware programmes and the use of VBI lines to download software directly to certain brands of microcomputers. Current applications include investment, education, and tax packages.

9.6 Advertising on teletext

The United Kingdom and Switzerland are the only European countries in which advertising has been developed and is permitted on teletext. At present in the United Kingdom, the general 'rule' arbitrarily defined is that there should be no more than 15% of the teletext magazine devoted to advertising.

Numerous advertising options are available from Oracle. The basic weekly rate for one page is $2150, dropping to $1500 per week for 40 weeks. An individual or company may insert a one-time advertisement or message for as little as $10. A year's subscription can result in discounts of up to 25% and purchases of more than one page can also result in large discounts. There are currently 30 advertisers who have national teletext advertising campaigns. ITV has announced a major regional teletext service for the United Kingdom. As with the national magazine, each 100-page 'regional' magazine is permitted to allocate 15% of the total pages to local advertising. Direct response offers, customer information services, and special sale offers are all being tested.

A key aspect of ITV's promotional campaign is the fact that one of teletext's strengths is its ability to *complement* existing television, radio, press, and direct marketing campaigns. Advertising is also being sold in company 'blocks' of pages. British Rail, for example, has signed a long-term multi-million dollar contract for 20 full pages of advertising and editorial information. It promotes a range of consumer services, including special offers, Railcards, intercity journey timetables, and standard product information. It provides automatic updating via computer link-up so that passengers will have up-to-the-minute information on departures and arrivals.

The relative purchasing power of a dollar of advertising expenditure using various media in the United Kingdom in 1985 is interesting (Table 9.4).

Although advertising rates are still nominal, evidence exists that companies are willing to make investments in advertising space on teletext data bases. In recent surveys, findings suggest that about 70% of all Oracle viewers were aware of teletext advertisements and 44% access advertisements through the advertisement index.[5] Further, there are 200 pages on a

Table 9.4   The relative cost of various kinds of advertising, UK in 1985

Medium	For an expenditure of $2000 you can buy:
Teletext	One full page (television screen) on a teletext magazine (cycling time for the whole magazine is about 30 seconds) for one week of broadcast transmission (about 960 characters; i.e., about 200 words); price does not include production costs
National newspapers	$\frac{1}{2}$ page × 2 columns in black and white in the *Sunday Times Magazine*, or 2 cm × 2 columns on the front page of *The Times*
Radio	Four 30-second spots in peak time for Capital Radio, a commercial station in London
Television	(a) Half an average 30-second peak spot on Channel 4, or about 1.5 seconds of a 30-second spot on national ITV
	(b) Four 30-second spots on Sky Channel at February 1985 prices

teletext magazine, 15% advertising pages, and around 200 000 teletext viewers per day. Then, if advertising is assumed to be 5% of the pages viewed and the average daily household use of Oracle is about three accesses—for a total of about 30 to 40 pages a day—then, on average, there are about 10 000 views of each advertising page. The fact that teletext advertising requires a constant choice by the householder indicates that the medium will be oriented less towards persuasion and familiarization and more towards information about a product. It has an obvious value-added dimension for television advertisements.

9.7   Teletext user profiles

The findings reflect similar use patterns for each of four countries sampled: the United Kingdom, Austria, the Netherlands, and West Germany. This is not really surprising because most teletext services are fairly similar in content. These findings were obtained from a series of telephone interviews with members of broadcasting associations in each of the countries.

Across the board, the average teletext viewing time per day, excluding subtitling, is estimated to be around 10 to 20 minutes, with people watching some 30 different teletext pages. Viewers tend to watch the same set of pages each time, indicating that very specific uses for teletext evolve fairly quickly (Table 9.5).

The viewing habits suggest that teletext is *not* entertainment, but rather a current information service. National surveys have identified the most frequently viewed topics on the various teletext magazines—news, weather, sports results, and television guides being of major interest. Viewed topics are listed in Table 9.6 in decreasing popularity for each country.

218

Table 9.5 User profiles of selected teletext systems[6]

United Kingdom	Austria	Netherlands	West Germany
Majority of males	Mostly males	High users of newspapers, radio, and television	Men outnumber women (4 to 1)
70% in the 15–44 age group; large over-65 population not yet heavy users	25–44	—	Younger than average
Affluent and television oriented (a heavy TV viewer is defined as one who spends more than 30 hours per week at the set; this group constitutes 50% of teletext users)	White collar workers The more 'intelligent' sectors	Above-average income Higher-than-average education	Above-average income Higher-than-average education

Table 9.6 Most popular teletext services[6]

Austria	Netherlands	Sweden
Weather	News	Indexes/headlines
News headlines	Weather and traffic	News
Television guides	Television guides	Television guides
	Sports	Consumer information
	News headlines	
	Sports	

United Kingdom		West Germany
BBC	ITV	
News ·	News	News
Sports	Weather	Weather
Indices	Television guides	Sports
News flashes	Sports	Television guides
Finance		Newspaper headlines for the next day
Weather		
Fun/puzzles		

9.8 Development of the European teletext market

A key development is the possibility of moving to *interactive teletext* and higher levels of graphics, such as the dynamically redefinable character sets (DRCS) and redefinable photographic quality displays. Interactive teletext points towards two possibilities: downloading software or information, a development that creates interactivity within the terminal, and hybrid telephone/broadcast possibilities that open a whole new domain for the telephone companies. Downloading is now quite common and, in the United Kingdom, ITV has approached the government to permit pay-teletext closed user groups as part of the general teletext service.

A second new development includes priority ratings for various pages to reduce their cycle time. As the magazines have increased in size, the most frequently requested pages such as indexes and news headlines are inserted in a cycle two and three times. Subtitling has also taken a high priority. In the United Kingdom, subtitling increased from 1 to 4 hours per week in 1981. In Sweden, subtitling averages 3 hours per week and live 'textual' commentaries, such as major sporting events or royal weddings, are an additional subtitling service. In West Germany, subtitling is around 3 hours per week.

The impending DBS and cable developments have resulted in programme providers developing plans to offer teletext services on their channels for programme details and advertising information for pan-European channels. This represents a move away from broadcast dominance of teletext. In the United Kingdom, for example, British Telecom offers a teletext magazine derived from its Prestel database for programme providers. Thorn EMI has also launched a group of teletext feeds for cable systems in the United Kingdom, e.g., Sky Channel's Sky Text service of programme guide, promotional material and some Prestel data. The teletext service contains information related to the video content of the respective channels as well as mini-magazines. In the United States, Dow Jones began experiments in 1984 to use the vertical blanking interval and the FM sidebands to communicate money and commodity market information.

A further proposed enhancement is the W. H. Smith Videoline Channel, which is a full cable channel of text and video commercials. This can lead into catalogue or at-home shopping, as well as revolutionizing the concept of advertising on television. It is the transactional capability that ultimately makes the service of real value to the consumer.

Teletext in France has not been directed at mass markets. Rather, teletext services have been developed for special audiences and on a fee-for-service basis. The stock exchange services, Antiope-Bourse, which began in 1977, broadcasts stock reports on over 2000 stocks to brokers and money managers. The services use 60 scan lines of the old 819-line television network that is being phased out in France. It has proposed to expand the service to about 300 lines, i.e., a sufficient bandwidth to support 5000 to 10 000 pages of broadcast teletext.

Other specialized teletext magazines include: Antiope-Postes, a service providing information to post office employees; Antiope-Meteo, a weather service for workers in transportation (aircraft and shipping), tourism, and agricultural industries; and Antiope-Orep, a general information magazine for those in the southwest region.

Teletext receivers now account for between 20 and 30% of the annual sales of television sets in Europe. This percentage will probably rise as prices decline further. It costs only \$40 to manufacture and include the decoder module in a television set.

In summary:

1. Teletext receivers have now achieved a critical mass of television sales in at least seven European countries.
2. Most systems are now using between four and six VBI lines and expanding the nominal size of the total magazine to at least 200 pages.
3. Database content is primarily a combination of 'radio news' and newspaper information. The development of subtitling has proved to be a valuable public service.
4. User profiles are remarkably similar among European countries. They indicate that the early buyers of the technology are relatively young (25 to 45), predominantly white collar workers, above-average viewers of television, and more likely to be men than women.
5. New media opportunities are developing. Pre-prepared teletext magazines are available from British Telecom and Thorn EMI, and programme providers are evaluating full-text and text/video information channels for cable systems and satellite channels, and closed user groups are being introduced.

9.9 Path of entry into households for information services: the home computer

The home computer markets in the United Kingdom, in particular, and, more recently, West Germany and France have begun to show promise of being major consumer markets throughout the rest of the eighties. Market statistics vary according to sources and industry forecasts are notably unreliable. It is estimated that at the end of 1984 household penetration reached:

United Kingdom	930 000
West Germany	150 000
France	120 000
Benelux	90 000
Scandinavia	75 000
Italy	45 000
Spain	30 000
Other	60 000
	1 500 000

with sales of $1.8 billion.[7] By 1987, household penetration is expected to be around 50% in the United Kingdom, 25% in West Germany, and 15% in France.[7]

In all markets the professional use of computers (PCs at work and workstations) is growing more rapidly than the home market and is expected to be the dominant component of total sales. That being said, home computers are still a rapid growth industry, estimated to be growing annually by 50 to 55%, but new demand is expected to taper off sharply and the replacement market, e.g., upgrading equipment, is not expected to be significant until there is a major technological innovation in the home PC industry. The market for peripherals—printers, floppy discs, hard discs, compact discs, colour monitors, and modems—is just beginning.

For the consumer, the choice for peripherals is clear—maximum flexibility, i.e., compatibility and upgradability, at minimum cost. The supplier, on the other hand, wants to maximize penetration; this can be achieved either by offering a comprehensive total system (at a higher investment cost) or by offering a less extensive array of devices and allowing the customer to shop around for specific peripherals.

The trend is towards modular design systems. Starting with a basic kit, comprising a processor, QWERTY keyboard, and videocassette drive, possibly with a joystick (priced at less than $15)—all operating on a television set—the common peripherals are:

Input/output	VDUs monochrome or colour, with high-resolution graphics.
Output	Printers—usually thermal or matrix dot impact printers for the household. Colour printers and graphics printers have not reached mass market prices as yet.
	Voice syntheizers are beginning to be offered.
Input	Plastic pressure pad keyboards as well as keypads, including the touch-tone telephone, joysticks, light pens, and touch-sensitive displays.
Storage	Audio cassettes, floppy disc drives ($5\frac{1}{4}$ inch) and the micro floppy discs (3 and $3\frac{1}{2}$ inch).
	Hard disc drives are still not commercial for the household, although compact discs are beginning to make inroads for high-quality music, graphics text still-frame and limited full-video motion.
Interface	PTT domination and strict control of the use of telephone lines for data transmission has slowed down the public access data-base market as well as the use of modems (the modulator demodulator necessary to convert a digital signal to analogue for carriage on the telephone

system). The introduction of public access packet-switched services and the deregulation of modem equipment (in the United Kingdom, France, and pressure now in West Germany) guarantees that, as in the United States, networking, messaging, and accessing of external databases will offer significant growth potential across Europe.

The UK market has achieved the highest per capita penetration of personal home-based computers in the world, due primarily to the Sinclair Z80 and its replacement the Z81. These machines were the catalyst for the industry as they offered the consumer a personal computer, albeit limited, for less than $150. The more 'advanced' of these machines, the Z81, arrived on the market with a usable memory (RAM) of 48 kilobytes, a 40-key keyboard enabling 200 functions to be performed, and its own user-definable graphics. The software supports were primarily electronic games capable of being loaded by commercial portable cassette and a modified version of the programming language BASIC.

Other early entrants into the PC market included Commodore, Acorn and Tandy as well as Atari, Texas Instruments, Dragon, Tycom and Camputers. Those in the latter group have either withdrawn from the market, refined their products, or simply gone broke.

By 1984, the home PC market was beginning to 'shake out', with growth rates down from 1983. Yet, the major suppliers of equipment for the home (priced at less than $600) were still offering non-compatible software. Commodore, Acorn/BBC, and Sinclair, who together dominate the home markets in Europe, all offer predominantly non-compatible software. The Japanese, in an attempt to enter this market, have developed a new software standard MSX. The standard is simply a common standard for 8-bit personal computers and programs developed by Microsoft with Japanese ASCII.[8] The standard is designed to be upward compatible and so far over 40 major manufacturers have agreed to use the standard. This gives Toshiba, Sony, Sanyo, JVC, and Canon a 'common marketing push'. However, it may be that the format has arrived too late to become the European standard, if there is to be one.

Although 1984/1985 was the year of the shake-out in the personal computer market, the number of machines installed and the more aggressive policies of the PTTs to open up the market for modems ensures a steady growth in this industry over the next 5 years.

9.10 Videotex services

It is now becoming clear that the personal computer is the path to home-

223

based enhanced electronic services such as electronic mail, transactions, telesoftware, teleshopping, telemonitoring, and two-way information retrieval. It remains to be seen whether videotex, as originally conceived, will survive, or whether the array of services which comprise a videotex system—information retrieval, electronic mail, transactions, computing, telesoftware, and telemonitoring—will evolve separately. France is the only country in Europe where there is a substantial videotex home market. It has a specific applications-based approach: an electronic telephone directory which is being introduced at subsidized rates by the PTT.

The personal computer is the predominant path of entry to the household for electronic information services. The growth rate and achieved levels of penetration have exceeded expectations. When video game consoles are included within the definition of home computers, the European market exceeds 7 million households or 6% of all households.[7] Videotex has not had near that level of acceptance anywhere. The national support for Teletel in France, the electronic telephone directory system, has ensured a significant growth in that market where current installations exceed 1 000 000, which is still lower than originally predicted by the French PTT. Elsewhere, the *total* European *home* market for videotex is less than 200 000 units, a clear indication that the 'right' combination of services, hardware, prices, and consumer receptivity has not as yet been found.

The videotex trials and services offer great promise but the limitations of a ubiquitous technology compared with specific technologies for specific applications seem to outweigh the benefits. Three significant observations can be made.

First, regardless of the purpose or original intention of videotex, the business user is now the prime initial focus in most instances. Videotex, therefore, competes with, or complements, a host of on-line database services targeted to their respective user markets.

Second, the specific applications (again, mostly business) will lead to videotex, *not* the converse.

Third, this relatively low level of penetration has taken place against a backdrop of limited competition from electronic mail, networking for home computers, or external data bases for home devices—competitive factors which are beginning to take shape in Europe. Further, time-sensitive costing for users, hierarchically structured menu designed access protocol, and limited value-added services such as transactions have all contributed to the low penetration.

The PTT domination of the network design and interface to the database has also contributed to the long lead time in establishing commercial services. Videotex will not be really market tested until after 1985, when the national experiments are replaced by market services. To this end, one can be reassured that the large investments by PTTs and their conservative approach to implementing new technology will mean that as decisions are taken to go ahead, there will be considerable clout behind these decisions.

Designs of national videotex systems are all unique, although we can classify the public access videotex systems according to some general characteristics (see 9.11). A videotex system requires an information or service provider, a network on which to communicate the information, a videotex centre for the system operator, databases to be accessed by users and updated by information providers, and user terminals.

This classification leads to three general database configurations across Europe.[2]

1. *A centralized database model*   This was the original Prestel design where British Telecom maintained a master database at a centralized update centre and a replicated database at each regional information-retrieval centre. The introduction of packet-switching (in 1980) is allowing the system to be modified to allow access to remote databases.

2. *A gateway design*   The Deutsche Bundespost, Bildschirmtext System, was designed as a gateway system. Originally based on the British Prestel format, the DBP announced in 1981 its intention to move to the CEPT standard. This path, while still the official policy, has fallen behind schedule as the decoder chips necessary for CEPT have proven extremely difficult to develop.

   The transport of data on Bildschirmtext uses Datex-P, the German public packet-switched network. There is a central Bildschirmtext database and access to external computers occurs via gateway pages in this database. The user accesses the Bildschirmtext database through the public switched telephone network.

   The prime motivating factor in the German design was the expected importance of catalogue shopping and reservation services, much of which is already computerized. Thus, the need for external databases, without centralized duplication, was seen to be a significant design factor. A similar model was developed in the Netherlands to run on Datanet 1, the national packet-switched network.

3. *Distributed gateway design*   The French Teletel system extends the distributed database concept one step further. The service is designed around Transpac, the national packet-switching network, with the French DGT's policy that the PTT is to act only as a 'carrier of information' at the core of the design. The link between a user terminal and any distant database is established by the user over the local telephone network to the nearest 'distributed gateway' and then via Transpac to the selected database.

A series of intelligent videotex concentrators on the network handle user passwords and provide menus of videotex services.

9.11 Videotex system status

Videotex systems offer an array of information, messaging, data processing, communication, and educational, entertainment directory, and transactional services. There are four general classes of videotex services which have evolved in Europe:

1. *Public videotex*   The systems have been developed by PTTs in Europe and are aimed at supplying stand-alone videotex terminals or videotex decoders for computers to the business and home markets.
2. *Public-access videotex*   These are networks of remote terminals (e.g., in shopping malls or public places) which may draw on information in the databases on the public videotex systems.
3. *Closed-user group videotex*   Again, the facility may be available on the public videotex system but the concept is one of limited access to information. A user needs to 'belong to a club' to be able to access the system, e.g., stockbrokers, medical personnel at a hospital, marketing representatives for a company, members of a bridge group.
4. *In-house private videotex systems*   These are videotex systems which are located within a single corporation or group of corporations but are not accessible to the general public. The system may be offered on a local area network, the public switched telephone network, or a national packet-switched service.

Our concern is with public videotex systems, those systems which are designed for the mass market rather than the business or professional markets. They may offer closed user group facilities and supply information to the public access systems.

Reviews of the status of videotex systems (including number of terminals, number of information providers, cost to store frames, and services available) appear regularly.[9] We have classified European nations into three groups reflecting the level of development of public videotex:

1. Commercial system operating
2. Trials under way prior to commercial service
3. No national videotex systems by 1985

9.11.1   COMMERCIAL PUBLIC VIDEOTEX SYSTEMS

This category includes countries with 'established' videotex systems, i.e., the trial or testing period has been completed and a commercial service is available (Table 9.7). The European countries with such systems are Finland, the Netherlands, Sweden, the United Kingdom, and West Germany.

While Prestel is the predominant standard, the long-term stated strategy of most countries is to move to a CEPT standard system. In France, the only country which seems likely to maintain a high penetration of terminals

Table 9.7 Commercial videotex systems[3, 10]

Country	Year commercial service began	Service name	Terminals* at end of 1985 (est.)
Finland	1980	Telsat*	2 000
France	1982	Teletel	1 000 000
Netherlands	1984†	Viditel	14 000
Sweden	1982	Datavision	6 000
United Kingdom	1979	Prestel	51 000
West Germany	1983	Bildschirmtext	14 000

*There are 12 private Telset systems; the first Helsinki Telset was introduced by Sanoma Corporation and began commercial service in 1980. The Finnish PTT also operates a system.
†Viditel began its commercial service in 1981, but it was not until 1984 that the Dutch government formally authorized the full introduction of videotex.

in the domestic market, there is no policy to move away from Antiope. Further, although most nations began with a centralized database design for videotex systems, the trend is towards gateway services.

With the exception of Finland, where regulatory problems prohibit the various telephone companies cooperating with each other and where the PTT has been slow to introduce gateway facilities, the gateway facility has been central to what growth there has been in the videotex market. On the French system, the PTT does not own or operate any databases except the electronic directory databases. All videotex services are accessible either through the public switched telephone network directly or via Transpac, the public packet-switched network. In Germany, the Bundespost not only provides gateway links to external databases on Bildschirmtext, but is also an information provider on the system. The Bildschirmtext design is such that the most frequently accessed frames are stored in local videotex centres. If the caller requires a frame not held at the local level, the call is automatically rerouted to a regional videotex centre, and if still unsuccessful, to the central videotex computer in Ulm that contains the whole database (Table 9.8).

Videotex services have not achieved profitability as yet, although some individual information providers have operated successful ventures. There has been a major policy shift in focus from households to businesses for most systems. In addition, numerous private systems are now in place— ranging from 375 in the United Kingdom to 25 in Finland. The array of information services, the relatively crude methods of retrieval, the limited information per frame, the 'unfriendliness' of VDU or television screens for 'reading' large amounts of information, the limited availability of interactive services such as banking, shopping, and bill messaging, and the cost of

Table 9.8 Structure of videotex systems

Country	System operator	Existing standard	Proposed long term	Database design
Finland	Individual companies*	Prestel	CEPT	Centralized DB for each system; no networking
France	PTT	Antiope	Antiope	Distributed
Netherlands	PTT	Prestel	CEPT	Decentralized; gateway facilities central to growth
Sweden	Televerkert	Enhanced Prestel	CEPT	Limited gateway architecture
United Kingdom	British Telecom	Prestel	CEPT	Centralized initially; Prestel gateway now operational
West Germany	Bundespost	Prestel/ CEPT†	CEPT	Decentralized‡

*The first, Helsinki Telset, was jointly operated by Sanoma, Nokia Electronics, and Helsinki Telephone Company.
† The objective was to move from the Prestel trial, which began in 1980, to the CEPT system in 1983. There was a 12-month delay in producing CEPT decoders and it was only during the latter half of 1984 that both systems operated. The phasing out of Prestel should be completed in 1985 as the German version of CEPT uses an 8-bit character code, making the Prestel terminals, which use a 7-bit code, incompatible.
‡ In 1985, 12 external computers were connected to the PTT gateway network.

the service have all militated against domestic *consumers* at this time. The rapid growth of teletext is a further dampener to videotex, real-time information (news, sports, weather, and television guide) being readily available at no ongoing cost.

The business focus is most predominant in Finland, the Netherlands, Sweden and the United Kingdom. In the Netherlands, for example, Viditel is targeted to specific industry sectors including travel agents, transport (especially freight forwarding), medical and pharmaceutical, agriculture, finance and banking (but not home banking), and educational and professional sectors who want to use the telesoftware service. This is similar to the United Kingdom, where British Telecom has developed a number of joint ventures to offer targeted services including: financial (Citiserve is a cooperative with the London Stock Exchange), farming (Farmlink is a cooperative venture with the Ministry of Agriculture, the Meat and Live-stock Commission, and the Meteorological Bureau), travel, banking (Homelink is a *home-based* facility provided as a joint venture with the Bank of Scotland and Nottingham Building Society), and telesoftware. The banking service is aimed specifically at the home market.

In addition, specialized videotex services are being developed for sectors such as real estate agents, doctors, insurance brokers, and educational establishments. The home shopping and electronic mail order shopping aspects of videotex are key elements in the design and believed demand for the service in Germany.

The most promising developments for the home market come from France. The videotex terminal, Minitel, a small monochrome screen ter-minal with built-in modem, is relatively inexpensive to buy or rent. This is a deliberate policy to encourage household penetration. The path-of-entry service, an electronic telephone directory (plus numerous other databases) with spelling checks, synonym choices, prompts, and text searching has been popular and has long-term economic implications for the PTT (up-dating a database versus preparing and distributing an annual paper-based directory). The PTT support for this service includes subsidizing the cost of terminals and not charging users or information providers for database access.

In France, the transaction side of videotex (teleshopping and telebanking) is being developed through a smart card, a plastic card with an imbedded microprocessor. The aim is to make smart-card readers available in public places and on the Minitel terminals. The French company Bull has formed a partnership with Mastercard in the United States to launch two electronic pay experiments in the United States. In addition, the Carte Bancaire Group (Crédit Agricole and Carte Bleue—French Visa) will issue 2.5 million smart cards for use in a programme in Brittany, the North, the Riviera, and the Rhône-Alps. The PTT has also ordered 1 million smart cards for various services from electronic phone booths to postal check electronic payment services.

In West Germany, on the other hand, electronic transactions—ordering, booking, and remitting—are being developed as gateway services on Bild-schirmtext. They are the most popular services, and the largest group of companies participating in Bildschirmtext are mail order companies, whole-salers, and retailers. Over half the orders placed are received outside normal business hours (09.00 to 18.00).

An electronic telephone directory service has been announced for Norway and West Germany, both to begin in 1986.

9.11.2 PUBLIC VIDEOTEX AT THE THRESHOLD
In this category, we include all countries in which a public videotex service

Table 9.9 Emerging videotex systems[4,11-13]

Country service/ name standard	Status	Focus of service	Number of terminals	Range of services
Austria Mupid/CEPT	Trial underway; operational service late 1985	Households	500	Information, electronic newspaper, sales catalogues, home computing
Belgium Bistel/Prestel	Trial until 1985	Business groups, households	1000	News, travel services, stock exchange, business information
Denmark Teledata/Prestel	Trial until 1985	Businesses/ households	1000	News, general information
Italy Videotel	Service coming in 1985	Business service and closed user groups for 12 months, then general public service 1986	1500	General and specialized information services, finance data, shopping/ transaction services
Norway Prestel	Trial until service in 1985 as a CEPT standard	Business first	150	Residential service proposed for 1986 with an electronic directory
Spain Ibertex/CEPT	Trial until operational service in 1986	Professional users in 1986	600	Business, shopping, travel, and education information; gateways to extend databases
Switzerland CEPT	Trial until operational service in 1986		3000	

was being tested in 1984 or for which an operational service commenced in late 1984 or 1985. These countries include:

Austria
Belgium
Denmark
Italy
Norway
Spain
Switzerland

While the array of services differ within countries, information retrieval and transactional capabilities dominate most of these trials (Table 9.9).

9.11.3 NO PUBLIC VIDEOTEX SYSTEMS

Those countries which have no videotex systems whatsoever include Portugal, Greece, and Ireland. In Greece and Ireland, videotex services are being tested for commercial user groups, but there are no immediate plans to develop public videotex systems.[10]

The systems which are about to come on-stream all indicate the change in direction for videotex. No longer is the general residential market the focus; targeted business applications are being developed as the first step before any roll-outs to the consumer markets.

9.12 Evolving videotex markets

The growth of videotex in Europe, if left solely to market forces as in the United Kingdom, will continue at a relatively slow pace in the home market. The consumer, who may want each of the services associated with videotex—information retrieval, transactions, computing, home management, and messaging—will tend to get them from specialized hardware (or software) systems.

This does not stop enthusiastic projections for this market. For example, Butler and Cox in their private (individual terminal) videotex forecasts suggest that by 1993 there could be 28 million home and business terminals in Europe with an annual spend on videotex hardware and software of $4.7 billion. They estimate the residential market is between 12 and 17 million terminals.[10]

Such forecasts are predicated on the French PTTs' continued support for the electronic directory and the French market's contribution of some 6 to 7 million home terminals; the continued support for Bildschirmtext and a household penetration level of 3 to 5 million terminals; the eventual take-off of videotex in the United Kingdom (which could include home messaging services being developed on Telecom Gold—the public packet-switched network), resulting in a residential market of 2 to 3 million terminals and a rest of Europe (Scandinavia and the Netherlands particularly) market of 1 to 2 million terminals.

Given that the French Teletel directory is the only service with over 100 000 subscribers, it is clear that to achieve the above levels of penetration will require continued decline in prices of hardware, improvements in services, and direct PTT support. Even a market of about 50% of the above appears optimistic.

9.13 Conclusion

The late eighties will determine whether there is a market niche for videotex or whether an array of home-delivered interactive services will come through specialized electronic developments (e.g., home banking) and online database corporations such as CompuServe and The Source, which have emerged in the personal computer market in the United States.

Teletext is assured of success in Europe, although it has taken on a more limited role than originally believed. Television sets are now routinely including teletext decoders and the enhanced graphics promised in the high-level protocols of CEPT, namely the dynamically redefinable character sets (DRCS), may increase the appeal for education, business, and the household.

Already the closed user group concept has been tested in France and suggested by the ITV companies to the IBA in the United Kingdom. Downloading of software has been demonstrated successfully and this opens the possibility of more extensive use of the VBI teletext services and even full-channel teletext in the non-broadcast hours. The continually falling price makes national messaging (targeted if closed user group facilities are in place) and electronic delivery of 'hard' information, technical realities. In the United States the numerous teletext trials have achieved virtually zero penetration primarily because teletext lacks broadcaster support. The full-channel scrolling text and graphics Cabletext services have been much more numerous, suggesting a potential, when and if new cable installations begin. Even so, the services are not popular with advertisers.

Cable system analyses in the United Kingdom have suggested that interactive services will be necessary for their economic viability, and this may provide additional avenues for distributing local, national, and international teletext and videotex services.[14] 1986 will be a crucial period for videotex as most major European nations have moved from the trial or experimental stage to a full commercial service.

Value-added services, particularly electronic messaging, on the public packet-switched network may stimulate the home market, especially if the price of modems continues to fall and PTTs support the notion of integrated modems in manufactured terminal equipment. Personal computers, already a sizeable growing market, will begin to take on enhanced capability, with a proliferation of peripherals. By the late eighties, networking, bulletin boards, and messaging will be commonplace, aided by lower prices for modems and their integration within the computer.

Butler and Cox, in assessing future trends in videotex systems, suggest four broad areas in which changes that are taking place will stimulate the videotex market.[10] These trends apply equally well to home computers or, generically, home management information systems: software, hardware, communications, and converging communication/computing technologies.

On the *software* side, the trends for home computers and videotex terminals are:

1. Keyword access systems becoming available for micros
2. Powerful electronic messaging facilities
3. Database management systems
4. Customized software development aids
5. Special videotex application packages such as telephone directory systems
6. Telesoftware libraries
7. Personal management and telemonitoring packages
8. Voice-activated systems

The Austrians are the first to move towards a policy of total user security when accessing a system. The policy allows a user to call the service anonymously. While offering protection, it also brings into question the whole concept of transactions on the system.[15]

On the *hardware* side, the general trends for the computing industry will continue, i.e., exponential declines in the cost of disc storage, computer memory, and computer processing power.[16] In addition, the trend is towards intelligent terminals suggesting either a rethink in the single-purpose videotex terminal or the inclusion of hardware (or software) within a microcomputer to provide a videotex capability.

On the *communication* side, the trends are more general, being associated with the evolution of the national telephone systems, and include:

1. High-quality data transmission rates of up to 50K bits/s
2. Increasingly intelligent networks to provide value-added services such as encryption, billing, error detection, protocol conversion, and service directories

The most significant development on the network side is the move towards European telecommunications. Through bi-lateral videotex network agreements, France and West Germany signed an agreement in late 1984 whereby the Teletel and German Bildschirmtext networks can be interconnected without regulatory obstacles, although tariffing policies are still to be

233

resolved: France charges an access time and West Germany charges on the number of frames accessed.[17] France is seeking a similar cooperative agreement with Italy, Switzerland, Belgium, and the Netherlands, thus linking Teletel, Prestel, and CEPT Standard networks.

Finally, the often-discussed area of integration: *convergence of telecommunications and computing* is shaping videotex and the home computer so that:

1. Videotex terminals contain an integrated telephone handset and voice capability
2. Videotex terminals become microcomputers
3. Interlinking of private and public data and voice networks is transparent to the end users

In summary, the videotex market will depend on the continued commitment of the PTTs. The electronic directory in France is an application-driven stimulus to that national market. The integration of videotex and the home computer, the accessibility of the public packet-switched networks, and the development of specialized cost-effective services for the household will ensure continued growth in Europe in this market.

References

1. Ogilvy and Mather Europe, *The New Media Review 1984*, Ogilvy and Mather Europe, London, 1984.
2. J. Tydeman, H. Lipinski, R. Adler, M. Nyhan, and L. Zwimpfer, *Teletext and Videotex in the United States*, McGraw-Hill, London, 1982.
3. 'Videotex statistics', *Videotex International*, 11 April 1985.
4. *International Videotex and Teletext News, International Update*, May 1984.
5. 'Audience research', Oracle Teletext Limited, July–December 1984.
6. R. Adler and J. Tydeman, *Teletext Market in the USA*, Institute for the Future, Menlo Park, California, 1982.
7. 'Computers in business', special section, *Financial Times*, 11 April 1983.
8. 'Home computers', *Financial Times*, 26 October 1984, pp. 13–17.
9. In such publications as *Videotex International* and *International Videotex and Teletext News*.
10. Butler and Cox, *Videotex in Europe*, Butler Cox and Partners Ltd, London, 1985.
11. *Videotex International*, November 1984.
12. 'Videotex in Europe', *On File Rapport*, November 1984.
13. *1984–1985 Videotex Directory*, Arlen Communications, Bethesda, Maryland, 1984.
14. 'Dialing up the profits', *Cable and Satellite Europe*, May 1984, pp. 21–23.
15. *Videotex International*, 4 January 1985.
16. L. Branscombe, 'Computer technology and the evolution of world communications', *Telecommunications Journal*, Volume 47, 1980, pp. 206–210.
17. *Videotex International*, 19 January 1985.

10

New media: A light in the regulatory tunnel

10.1 Introduction

In Chapters 1 to 9 we have reviewed consumer electronic media in Europe in all its uncertainty. Our intention has been to examine traditional broadcasting, cinema, and advertising (Chapters 2, 3, and 4) and describe new developments and potential markets in the modes of delivering entertainment to the household, namely, satellite (Chapter 5), cable (Chapter 6), and video cassettes (Chapter 7), as well as identifying the trends among the plethora of proposed entertainment channels (Chapter 8) and electronic information services (Chapter 9).

The message which emerges is that the euphoria of the early eighties has been replaced by the frustration, uncertainty, and difficulties of implementing major new communications projects within and across national borders. Optimistic projections for penetration have been scaled down by as much as 50% within 2 years.[1] Still, the milestones in Europe from the launch of Eutelsat 1-F1 to the legal reception of satellite-delivered programmes containing advertising in Norway have been substantial.

Frustrations with new opportunities aside, the evidence still points to a European television market which lacks variety of entertainment programming (Chapters 1 and 2) and commercial opportunities for advertisers (Chapter 3). The rapid growth of video cassettes, in many countries, is an indication of the potential demand for more and varied at-home entertainment. Prior to Eutelsat 1-F1, it was largely a matter of government broadcast policy that determined the broadcasting mix in each country. That is changing now as, little by little, cable systems begin to retransmit 'foreign'-originated programmes.

To conclude this assessment of new electronic media in Europe, we consider the key future policy and regulatory events which will shape (both enhance and inhibit) the growth of these markets. It is clear that there is still no universal European communications or media policy. Further, there has been no indication that a single nation can integrate the policy implications of satellite, cable, and other new media within a coherent policy framework.

Because broadcasting and telecommunications have been traditional public monopolies, and because governments in Europe are relatively cautious about market deregulation, a 'let-the-market' decide approach has not been forthcoming. Yet, European nations are making the greatest changes to their existing broadcasting frameworks and the trend across Europe is towards increased private enterprise participation in all aspects of new media.

10.2   Regulatory framework for new media

Eutelsat represents a watershed in the development of international broadcasting in Europe. Prior to the launch of 1-F1, international broadcasting in Europe comprised special European Broadcasting Union events such as news services, the Eurovision Song Contest, shared 'satellite' services, including the World Cup and other sporting and national events. These were all authorized through Intelsat satellites with the full agreement (and usually participation) of national broadcasters. There were no problems with downlinks but occasionally a sponsored event would break a national advertising standard, e.g., a cigarette company sponsored event, and occasionally there would be retaliatory actions as in the case of the Swiss banning the coverage of a World Cup skiing event held in Switzerland because the course was decked with billboards from the sponsor, a cigarette company.

There was also the 'intentional' spillover services such as RTL and Télé-Monte Carlo, but they were reasonably well contained. This position has changed significantly with the delivery of pan-European channels by non-national broadcast organizations.

In its report, *Realities and Tendencies in European Television*, the Commission of the European Communities outlined clearly the coming position of new media in Europe:[2]

> The new telecommunications technologies, and especially direct broadcasting by satellites, will inevitably lead to a proliferation and an internationalisation of television broadcasts in Europe by the end of the decade.... Programmes will inevitably become more international and this will directly affect cultural developments for the people of Europe and hence the future of the community.

They also foresaw the paradox that international television poses. On the one hand, it is an 'opportunity not to be missed for spreading a particular culture or message more widely' while, on the other, 'it poses a threat of invasion by a foreign culture'. They were firm, however, in their view on

236

the right of the individual to international choice: 'In theory a government could prohibit its citizens from watching foreign transmissions, although this would be unthinkable in the Western Democracies.' They also pointed out that the World Administrative Radio Conference of the International Telecommunications Union did not allocate any frequencies for a pan-European service and that, due to the advances in receiver antenna technology, the size of the audience outside any national market, the subject of considerable concern in 1977, was now larger than the national target market itself. This is the case with the French DBS service which has been presented to European programme providers as a European service! (see Chapter 5).

Direct broadcast satellite services will now follow satellite-delivered cable television and master antenna television into European nations. Fixed service satellites, with their 'point-to-point' transmissions, are fast becoming the forerunners to DBS through their potential to serve MATV households. The household composition in Europe is more conducive to master antenna systems than to individual reception. This suggests that a medium-power satellite system could be the optimal solution for most entertainment needs. Such a solution was proposed by Luxembourg and later by the Eutelsat Council.

The rules are being set by the early pan-European services, such as Sky Channel, Music Box, and TV-5. One way to view the regulatory context for new media is to trace the necessary steps involved to set up and operate a pan-European advertiser-supported television channel delivering its programme by satellite. It is assumed that the satellite system for such a venture is a 'recognized' European communications satellite (ECS).

Logically, the channel provider (the term we will use throughout to refer to the organization responsible for programming and having a television channel delivered by satellite) is faced with two sets of decisions: those determining the delivery of the service and those influencing the content. The delivery or transmission of a pan-European service involves three steps:

1. The lease and operation of the satellite system
2. The downlinking of the service internationally
3. The distribution of the service by national cable operators

On the content side the channel provider has two basic regulatory/policy concerns, in addition to the obvious economic and social decisions influencing the choice of content. They are:

4. The acquisition of programme product
5. The transmission of advertisements within the service

This structure is not perfect. For example, steps 2 and 3 are not completely independent as cable operators need a licence to carry an international channel in addition to the channel provider obtaining government approval to downlink. However, they do demonstrate the general procedure emerging in Europe to establish a pan-European television service.

10.3 Lease and operation of the satellite system

The arrangements for the uplink and transmission of a satellite service in most European countries are relatively straightforward. The national PTT is the signatory for both Eutelsat and Intelsat satellites. Transponders are leased by Eutelsat (or Intelsat) to the PTT who in turn leases to the channel provider. In the United Kingdom, where both Mercury and British Telecom are authorized under the Department of Trade and Industry to offer uplinking, the customer has a limited choice.

The channel provider must satisfy the Eutelsat requirements on encryption, which in 1985 merely required the channel provider to warn against illegal or unauthorized viewing of the service. The channel provider is also responsible for content (copyright clearance, libel, and conforming to advertising agreements). The PTT would generally be indemnified against the channel provider.

Eutelsat has specified a pricing formula, depending upon whether the transponder is eclipse-protected (or not) or whether the service is preemptible or non-preemptible and whether the channel provider wants to downlink nationally (to the country in which the signal originates) or internationally. In the latter case downlinks are 'sold' as extras on the base price for the transponder.

The PTTs, in turn, determine a cost-plus basis for lease to end-use lessees. All contracts to date by PTTs have specified that the lease is for a service rather than a transponder, e.g., to provide television, radio, or teletext, but not for the right to use the transponder for any (legal) telecommunications or transmission service.

In all European countries, the PTTs own and maintain the earth stations for uplinking to the Intelsat and Eutelsat satellites. The PTT control of uplinking follows directly from the fact that FSS services are defined under the national Telecommunications Acts, PTTs have a total national monopoly in this area, and the PTTs are the signatories to the Eutelsat and Intelsat agreements so they are in some sense the 'satellite owners'.

10.4 Downlinking a pan-European service

Services carried on communications satellites—the fixed satellite services (FSS)—generally fall outside national broadcast regulations (Chapter 2). The process by which a multi-national programme provider such as Sky Channel, Music Box, or TV-5 obtains permission to downlink the service to be received in a foreign country, differs from country to country. One element is common; it is the programme provider's responsibility to negotiate agreements with the respective national PTTs, not the responsibility of the PTT in the originating country. In general, the PTT in the originating country acts as the liaison between the channel provider and the other PTTs. This process has turned out to be easier for the channels on the Eutelsat satellites than those on Intelsat. It is important to stress that we are only referring to national approval to downlink, *not* to the process whereby cable operators choose to carry (or not carry) the service.

The case for satellite-delivered programmes to master antenna systems and for individual reception is quite different. As discussed in Chapter 6, the general thrust of the new regulations emerging is to allow the household, hotel, hospital, club, or MATV system to receive programmes from whatever satellite is chosen, subject usually to the acquisition of a licence.

While not all of the national new media laws have been established, most European countries receive one or more of the Eutelsat 1-F1 services, even if only on an experimental or trial basis. By December 1985, only Belgium, Greece, Portugal, Ireland, and Spain were not receiving at least one service. In Spain, most large hotels advertise the availability of Eutelsat programmes (Sky Channel and Music Box), although legal reception is still awaiting approval.

There are two general considerations in comparing the situation among European countries:

1. Whether the programme provider needs a formal agreement/licence to downlink
2. If there are restrictions imposed by the downlink country which do not necessarily exist in the country from which the service originates

A summary position is given in Table 10.1.

The French and Dutch, for example, require foreign programme providers to agree to quite different conditions. In the Netherlands, the requirement is that foreign broadcasters do not direct their advertising towards the Dutch. A licence is not required but a foreign channel must:

1. Be available in the country of origin in the same form as transmitted into the Netherlands
2. Have no advertising addressed directly at Dutch viewers
3. Include no Dutch subtitling in either programme or advertising material

On the other hand, in France, the concern is more for a quid pro quo agreement. The cable television regulation limits foreign programmes to a maximum of one-third of all programming on a given local cable network and, where practicable, the French seek reciprocal downlink rights with the foreign country.

Table 10.1 Downlinks for FSS services into European countries—requirements for Cable, December 1985

Country	Licence for programme provider	Advertising constraints on programme provider	Programming
Austria	No; experimental status	Nil	Nil
Belgium	Policy not announced	Nil	Nil
Denmark	No	Nil	Nil
Finland	No	Nil	Nil
France	Yes	Nil	French component; maximum non-French language programming per cable system
Greece	—	—	—
Ireland	—	—	—
Italy	—	—	—
Netherlands	Yes	Not directed at Dutch market	Nil
Norway	Yes	Nil	Nil
Portugal	—	—	—
Spain	—	—	—
Sweden	Yes	Nil	Nil
Switzerland	Yes	Block advertising; product restrictions	Nil
United Kingdom	No	Nil	Nil
West Germany	No	Nil	Nil

The position for those transmitting a channel on a direct broadcast satellite is more liberal, although DBS services do not begin until TDF-1 is operational in mid 1986 or later. The general understanding of the WARC '77 agreement and the subsequent resolution by the EEC and Council of Europe is that DBS *is* broadcasting and is national in focus. While the latter comment may only be true in spirit, the general climate for DBS seems to be more akin to the present broadcast spillover services than to one of foreign 'intrusion'.

There is one interesting phenomenon for the viewer: two national DBS satellites in different orbital positions. The restrictions in this case are technical, not regulatory. First, it is possible that the services could use non-compatible transmission standards. Second, different encryption systems may be in place on the satellite system. As a result, the receiving installation (household or cable system) may require two different antennas and two different decoders.

10.5   Distribution of the service across cable and SMATV systems

Once government approval to downlink has been received (if necessary), the cable operator's decision to take the service is based primarily on economic criteria (Chapter 6). Prior to Eutelsat 1-F1, the only channels that cable systems carried were national and contiguous-national broadcast channels which were intentional or unintentional 'spillover' services. Must-carry channels were usually defined to include all national channels. The arrival of DBS will increase the number of must-carrys for many cable operators and put pressure on the relatively small number of free channels available on most systems. The trend will be for national DBS channels to be must-carrys in their local country.

The extent of this must-carry trend is unclear. For example, France is looking to make TDF-1 into a European service and not just a French national programme. However, they would not necessarily grant foreign programmes carried on TDF-1 'must-carry' status in France. Will the West Germans feel obliged to include the TDF services as must-carrys on West German cable systems and will governments include SMATV systems, with their limited channel capacity, within the DBS must-carry obligations? In the interim, the FSS channel providers are competing on content and cost of their services for the limited availability on the systems. The programme providers are trying to straddle both market opportunities, and the evolution of syndication of programme services is sure to occur within Europe by 1986. Sky Channel has already begun to sell programmes to terrestrial broadcasters such as Télé Monte Carlo.

A restriction on the provider is that those receiving the service may need a licence. This position has not been clarified across Europe, but it seems likely that most major installations (i.e., cable systems) receiving FSS signals will need a licence before being able to distribute the services to cabled households (Table 10.2). The position for SMATV, hotels, and individual households differs significantly. In the United Kingdom, for example, all dwelling units installing an antenna will require a licence, whereas, in Sweden, only systems with 50 or more households need to be licensed.

The position of who-pays-whom—the cable operator for the programmes or the channel provider for 'access' to the market—is strictly one of negotiation. In Belgium, for example, a condition of entry for Sky Channel and Music Box was that they provided Belgian-made programming. For channel providers, a key issue concerns copyright protection and maintenance of station identity. To that end, the most legitimate concern is to 'control'

Table 10.2 Licensing requirements for earth stations—Europe

Country	Licence for individual reception	Licence for FSS, cable, and SMATV	Who issues licence	Ownership of receivers	Comments
Austria	No			PTT	Decision under review
Belgium	Yes	Yes	PTT	PTT	Decision under review
Denmark	Yes	Yes	PTT		
Finland	No	Yes		Private	
France	No	Yes	PTT	PTT	Requires PTT and TDF approval to start a cable system; PTT installs receivers
Netherlands	No	Yes		Private	Cable operator licence through PTT (over 100 connections); services licensed by local municipalities
Norway	No	Yes	PTT	Private	Not for systems of less than 50 households
Sweden	No	Yes	PTT	Private	
Switzerland	Yes	Yes	PTT	PTT	All installations by PTT; no licences needed by SMATV operators
United Kingdom	Yes	Yes	Home Office	Cable franchises	SMATV licences available outside of cable franchises
West Germany	No	Yes	PTT	PTT	Legislation pending

any retransmission by cable operators and to guarantee that the programme and its advertisements are transmitted intact. However, the cable operator is technically responsible for the content of the channel within the local country and consequently would usually be indemnified against the channel provider.

10.6   Acquisition of programming material

The acquisition of programmes, for any broadcaster, raises the fundamental issue of rights or copyright. Traditionally, European broadcasters have acquired broadcast television national language rights for their national markets. Beta-Taurus, a German film distributor, has been an exception. Beta-Taurus usually buys German language and original language rights for the whole of the German-speaking territories. Contracts may be for multiple or single showings, usually over a 2- to 5-year licence period.

The position for a new pan-European channel, however, is more complex. The programme rights matrix could include:

1. National, European, or partial-European market
2. Language rights (English, German, French, Italian, etc.)
3. Cable, SMATV, DBS, or other (e.g., video)
4. Pay or advertiser supported

Most major film distributors now accept this position for pan-European, non-broadcast channels and sell accordingly. It may not be possible to maintain this artificial barrier to the distribution of product across Europe indefinitely. First, the technologial developments with satellite transmission allow simultaneous language transmissions. Second, the pan-European advertiser supported market may be of sufficient volume to justify a change in the release pattern on economic criteria alone. And third, unlike the USA, the pay-television window may not eventuate to any significant degree in Europe.

The acquisition of the rights to film and television products does not cover the rights for the sound tracks as all broadcast performing rights for music are assigned to a national society (performing rights society) to act as the collection agency. In theory, the clearance of music rights is the responsibility of the cable operator but this clearance is usually negotiated in collaboration with the channel provider. There is no general European agreement across the performing rights societies and consequently music is still cleared country by country. PR societies have tended not to apply the same rules to satellite-delivered cable services as to national broadcasters, and have usually been prepared to deal with foreign-originated services more readily than local ones. The general agreement is for the channel provider to pay a local fee based on an audited audience size in the respective country.

The Nordic countries are considering extending the copyright clearance for music to other professions associated with television and broadcast products generally, such as for authors, directors, artists. This move, if implemented across the nations of Europe, would increase both the cost and the confusion for a channel provider.

10.7 Carrying advertising

We have already discussed the complex position for advertising on national broadcast television channels (Chapter 3). The Council of Europe, through its Committee of Ministers to Member States on Principles of Television Advertising (23 February 1984 and 12 October 1984)[3] produced recommendations for television advertising, especially when transmitted by satellite. The principles suggest guidelines for content, form, and presentation of advertisements. Five of the principles concern themselves with good taste, offering positive suggestions, e.g., no subliminal advertisements, and fair, honest, truthful, and decent advertisements. These exist for broadcasters now, although there are some fairly minor differences in the council's principles.

The other five principles, however, if applied across the countries of Europe, could shape pan-European advertising. Two relate to content and three to form and presentation. First, content:

> *A.* Advertisers should comply with the law applicable in the country of transmission and, depending on the proportion of the audience which is in another country, should take due account of the law in that country.

The implications of this recommendation are far-reaching, the most interesting example being Sky Channel.

Throughout late 1984/early 1985, in excess of 70% of Sky Channel's viewing audience was located in the Netherlands. The Dutch in their proposed New Media Law (which is being challenged in the European Court) have introduced specific provisions for overseas advertisers. Given their national position on public broadcasting as a part of the cultural state, their desire to keep the advertising mix between print and broadcast media relatively constant, the new legislation (which it is hoped will come into force in late 1986) restricts advertisements from foreign countries, forbidding them:

1. To be spoken in Dutch
2. To give prices in Dutch currency
3. To give addresses in the Netherlands
4. To include products only available in the Netherlands

But this regulation has not been well received. The European Economic Commission has challenged it on the grounds that it is in breach of Articles

244

59 and 62 of the Treaty of Rome which state that restrictions on freedom to provide services within the Community shall be progressively abolished (59) and that no new restrictions on the freedom to provide services shall be introduced (62).[4] The choice seems to be one of maintaining the status quo for advertising in Dutch media and watching the revenues flow out to Sky Channel, Music Box and other non-Dutch programmers, modifying the stance taken on the existing media or finding ways to recoup some of the advertising dollars from the foreign-originated programmes.

On the other hand, Norway and Sweden allow Sky Channel in its entirety to be relayed on cable systems, although neither country permits its own television broadcasters to advertise.

B. Utmost attention should be given to the possible harmful consequences that might result from advertisements concerning tobacco, alcohol, pharmaceutical products and medical treatments and to the possibility of limiting or even prohibiting advertisements in these fields.

There is no general broadcast position on the above products (Chapter 3). Worldwide trends have been to prohibit or at least limit advertising in these areas, although their revenue-generating power, where allowed, is substantial.

On the form and presentation issues, the principles are unequivocal:

C. Advertisements, whatever their form, should always be clearly identifiable as such.

D. Advertisements should be clearly separated from programmes; neither advertisements nor the interest of advertisers should influence programme content in any way.

E. Advertisements should preferably be grouped and scheduled in such a way as to avoid prejudice to the integrity and value of programmes or their natural continuity.

These recommendations are more stringent than the IBA Code of Practice in the United Kingdom. If interpreted precisely, they prohibit most sponsorship, and require the programme provider to transmit advertisements in blocks around the natural breaks in programmes. Even sponsoring special events would seem to be outside the spirit of these requirements, unless credits were limited to the beginning and end of the programmes. Yet, in practical terms, most big sporting, entertainment, and even cultural events could not take place without some degree of overt corporate sponsorship, an observation not lost on channel providers and potential sponsors alike.

As an example, the agreement reached by Sky Channel and the Swiss Federal Communications Ministry resulted in Sky Channel carrying in 1984:

1. Block advertising so that programmes were not interrupted
2. No advertisements for alcohol, tobacco, or pharmaceutical products
3. No sponsorship of programmes (although sponsorship of special 'live' events were allowed)

These restrictions were not as stringent as those applied to Swiss broadcasters who are not permitted to advertise on Sundays or insert advertisements between any programmes.

The Council of Europe's recommendations are a first step towards defining pragmatic guidelines for pan-European advertisers. Strict interpretation could mean an operational framework not unlike that existing in the German-speaking territories. Actual interpretation is likely to result in a code of practice along the lines of the IBA Code in the United Kingdom.

Thus, the signs are positive for new satellite channels in Europe. Mass media advertising is here to stay. The difficulty for the channel is to secure sufficient market penetration across European countries to provide the advertiser with a worthwhile market. The difficulty for member nations is to realize that irrespective of their national attitude to broadcasting (e.g., as a public service), 'foreign' advertiser-based channels will ultimately redistribute their national advertising revenue across national borders.

The objective is to devise a set of workable rules or, as is happening now, let the rules emerge in the market-place, to include such items as maximum transmission time devoted to advertising, sponsorship, separation of programming and advertising, prohibition and restrictions of certain products, as well as an attitude to children. The intention is to move towards a minimal set of guidelines for pan-European advertising.

10.8 Conclusion

New media markets in Europe will be relatively slow to evolve but the change to date has been significant. In 2 years, 1983 to 1985, some 18 new television channels began to operate in national and international markets in Europe. This represents an increase of almost 40% on the number of national broadcast networks. The telecommunication and broadcasting regulatory frameworks support the national monopolistic institutions and have militated against rapid change. New media legislation in European countries is creating a market opportunity on the one hand while maintaining the established state broadcasting system on the other. The role of advertising which historically has had a relatively minor and constrained role on the national broadcast media is particularly critical as neither governments nor subscribers have shown a willingness to pay the full cost of cable and satellite.

The entrenched regulated broadcasting environment has not constrained all new media growth. If anything, it is the state control of broadcasting which has allowed national teletext services to thrive; Europe is the only stronghold of teletext in the world and teletext is fast becoming a basic 'add-on' to colour television receivers.

Further, the lack of choice and limited fare of entertainment programming in Europe is one of the major factors why video cassette penetrations have been so high.

However, services which compete with broadcasters or require PTTs to loosen their stranglehold on consumer telecommunication products (such as modems, telephones, antennas, decoders) have not been as successful. Cable and satellite (either for cable relay, SMATV, or DBS) development have shown little of the rapid penetrations and industry growth witnessed in the United States in the seventies. All new cable and satellite programmes have fallen behind schedule, been reshaped to better equate to market realities, and had their pay-back periods extended. The significance of master antenna in Europe is beginning to emerge in 1985.

That being said, the market opportunity still exists—it is just 3 to 5 years later than predicted and the corporate players are less bullish as a consequence. The mix of cable, SMATV, and DBS will give many European households an additional four to ten channels, most of which will not be traditional broadcast channels. Advertisers are beginning to recognize that they have new options in creating their media campaigns and programme distributors see a variety of new rights windows in the release cycle of films, series, and musicals. For equipment manufacturers, SMATV and DBS offer a market opportunity and a technological challenge.

The newly formed cable channels, whether national or pan-European, are constrained only by government tardiness in granting downlinks or the physical capacity limitations of the old cable systems. Sky Channel, for example, almost single-handedly is setting the procedure for pan-European advertiser-supported broadcasting.

The most encouraging signs that Europe is at least prepared to concede the freedom of information-flow policy espoused in the founding EEC Treaty of Rome are coming from the Council of Europe and the Commission itself. Both bodies, while still adhering to the need for advertising guidelines and recognizing national sovereignty in policy making, are adamant in their support of the freedom to broadcast across frontiers. The EEC Green Paper, *Television Without Frontiers*,[5] encapsulates the basic conditions necessary for advertiser-supported television as seen by the Commission:

1. *Recognition of freedom of flow of information* In Paragraph 12, 'The Commission considers that all restrictions on freedom to broadcast across the frontiers of the Member States, whether discriminatory or

otherwise, are contrary to Articles 59 and 62 of the Treaty (of Rome).'
The right to pan-European or regional television is unequivocal.

2. *Recognition of sovereign rights of the programme originator*   While acknowledging the need for social responsibility in transborder broadcasting, the Commission states in Paragraph 13 that, in its view, 'These exceptions (such as public security, public health) do *not* [our emphasis] permit the Member States to apply national rules generally to the form, content and arrangement of programmes coming from other Member States whether by direct broadcasting or by re-transmission by cable of the foreign signal.' This again is a clear statement. Provided the programming satisfies the requirements of the originating country, there should be no other national restrictions on the programming.

3. *Responsibility by programme provider with respect to advertising*   The Commission acknowledges explicitly the shortage of available television advertising opportunities in Paragraph 24: 'In some national markets, demand for advertising time has been as much as double the available supply.' But, it cautions restraint (Paragraph 22): 'The Community would ensure that broadcast advertising would be able to make its positive contribution in all Member States, subject, of course, to further conditions. . . .'

Nations have been a little more wary. Most Governments realize that, historically and culturally, Europe is not equipped to be self-sufficient in entertainment programming, even if the problems and costs of dubbing were overcome. A start has been made. Olympus (Europa TV) represents a consortium of broadcasters producing a multi-lingual public service programming effort (Chapter 8), six television broadcasters have formed a production venture (Chapter 4); Music Box is undertaking joint UK/European production ventures and Sky Channel is doing likewise in its music programming. The outcome will be product for European and export markets. In the case of music, the UK is the dominant producer of popular music for western markets, a point not lost on the Americans. For 12 consecutive weeks in mid-1985 (May through July) UK bands held the coveted number one on the US pop charts.

Still, an open-doors policy for satellite channels almost certainly means an increase in the amount of American entertainment programming for European households. Given the importance of television in helping form cultural and societal values, it is fair to say that many nations see pan-European channels (or even more national channels) as part of an American-homogenization process. Local content rules, such as those proposed by the French, are one way to ensure a balance in the content available to households. The use of multiple-audio channels with video products and the emergence of worldwide co-productions may afford

248

European programmers the chance to increase their own markets and to achieve respectable penetrations of English-language markets.

In conclusion, there are some generalizations of the trends in European new media markets over the next five years:

1. Direct broadcasting by satellite, as defined in WARC '77, is still uncertain because of technological change, cost, the established position of low-power satellites, and the challenge of medium-power satellites.
2. Direct broadcast satellites that are launched, and TDF-1 is the most likely as it has the commitment and financial support of the French government, are necessarily considering the total 'catchment' population under their footprints and not just their national markets.
3. New cable systems are too expensive. National plans for fibre optic star networks are not economically viable when the cost of cabling (especially as most nations require the cable to be underground and not strung between telephone poles or other above-ground utilities) is linked with reasonable levels of household penetration.
4. SMATV from low-power satellites is technologically possible and provides an economic path of entry for new channels to apartment blocks, flats, terrace houses, and hotels and business establishments. However, the distribution of both antennas and programme packages are not at all certain and there still seems to be a need for a medium-power satellite system to maximize the potential of this market.
5. The mix of channels being offered will be entertainment-led with targeted narrowcast channels following. The 16 to 24 year olds, who have been badly catered for by broadcasters, are receiving considerable attention on the early channels.
6. Government broadcasters are taking a position in cable and satellite, thereby hedging their bets. Their contributions in the first 2 years have usually been either to rebroadcast existing material or to plan new channels using their existing philosophies (of cost, content, and quality).
7. New channels will only survive if they cover their costs through subscriptions (pay-channels) or, if advertiser supported, can be programmed at a fraction of the cost that the existing broadcast channels incur. Without full coverage of national markets, the case for advertisers is harder to make and takes longer to become established. An exception to this is the 'publishers' channel in Germany, SAT-1, where the programming costs have been allocated across a number of publishers (some 164 in all, although much of the responsibility falls on the large publishers). The criteria for profitability do not change, but the risk for any one company is reduced.
8. Publishers have been the first to attempt to find paths of entry into the television business through cable and satellite. In Germany, it is Ber-

telsman, with Gruner and Jahr who has formed a partnership with CLT, Bertelsman and Axel Springer, who have formed a partnership with Beta-Taurus to provide pay-television, and the publishers generally who constitute the members of SAT-1. In France, the Hersant Group (*Le Figaro*) and Hachette Publishing (with Filipacchi) who are trying to secure new terrestrial networks, and in the United Kingdom, Rupert Murdoch's News Corporation owns Sky Channel and Robert Maxwell, owner of the Mirror Group and Pergamon Press, has bought into both cable (through his holding in Rediffusion) and television (through his ownership of Mirrorvision).

9. Advertising outlets on the traditional broadcast media are in short supply across Europe. Pan-European advertising will emerge as a key new industry. Trends in recent years have been towards common European brand names within corporations. More importantly, cross-ownerships, joint ventures and takeovers, and multi-national corporate expansions have created an extensive set of international goods and services. The targeting of individuals in real time across national boundaries may create new advertising groups altogether.

10. National regulations will differ for advertising but through the EEC, Council of Europe, and EBU, an acceptable and workable framework for international channels is emerging. Nations trying to maintain a balanced distribution of advertising across all media (television, press, radio) will find this artificial balance more and more precarious as the number of foreign channels and their national penetrations increase.

11. Video cassette recorders are approaching a penetration plateau in the mature markets (the United Kingdom, West Germany, and the Netherlands) but are still growing rapidly in France and Scandinavia.

12. Teletext is established as an add-on to television sets in Europe. Interactive information services including videotex will generally follow entertainment into the household. The delay in some national markets is likely to be years as the home-based services such as banking and telemonitoring are expensive and require a degree of consumer awareness not necessary in entertainment services. The services, at least those available now, are often perceived to be of marginal value to the consumer.

Cable, DBS, or videotex will succeed or fail due to its economic utility to households. The European consumer has demonstrated a demand for more variety in the daily menu of television programmes. But like every good or service, there is a limit to how much the consumer is prepared to pay. The opportunity will be realized, but only if governments permit a regulatory context in which the industries can grow, and hardware manufacturers and programmers are prepared to take the subsequent market risks. The advertisers and consumers of Europe are captive audiences waiting for the opportunity. However, it will take attractive pricing, imaginative programming, and creative marketing before the real potential of this market is realized.

References

1. Conference Proceedings, 'Cable TV communications in Western Europe 1984', Communications Information Technology, London, Howard Hotel, 29 March 1984.
2. Commission of the European Communities, *Realities and Tendencies in European Television: Perspectives and Opinions*, Commission of the European Communities, Brussels, 1983.
3. Proceedings of the Council of Europe, Committee of Ministers to Member States on 'Principles of television advertising', Brussels, 23 February and 12 October 1984.
4. Commission of the European Communities, *Treaties Establishing the European Communities*, Articles 59 and 62, Commission of the European Communities, Brussels, 1984.
5. Commission of the European Communities, *Television Without Frontiers*, Commission of the European Communities, Brussels, 1984.

Index

255

259